An Introduction To A Study of The Stars

by

C.R. Mirbt

Clent Farm, Clents Grove
Stourbridge, Worcestershire
England

The Association of Waldorf Schools of North America

Published by: The Publications Committee
A.W.S.N.A.
3911 Bannister Road
Fair Oaks, CA 95628

Mitchell, David and John Petering, editors

Transcription by Nora Poole
Proofreading by Carol Wilson, LaHoma Stang, & John Petering

For use by the English-speaking Waldorf Teaching Community

AN INTRODUCTION TO THE STUDY OF THE STARS

ISBN # 0-9623978-5-7

Printed in the United States of America by Thompson & Shore Inc.

TABLE OF CONTENTS

Preface

While looking through his library for books on Astronomy in order to help a seventh grade Waldorf Teacher prepare her lessons, John found this little gem by C.R. Mirbt tucked way back in the bookshelf. Its rediscovery has helped many teachers and we wish to make it available to a wider audience.

Writing in England at Clent in Stourbridge in 1938, Mirbt brought out these studies originally in a newsletter format. Gathered here are ten chapters from this newsletter dealing with introducing geocentric Astronomy, which the editors believe will be very useful for teachers introducing the topic.

In the first three chapters the stars are mentioned only generally. The author takes his start from the human being standing firmly on the Earth. He believes that the more firmly we can establish our position here on earth, the more secure we shall be when we study the star's movements.

The second chapter begins with these star movements and then he explains how to use J. Schultz's Movable Planisphere. Next, he leads us in an exploration of the sky from the perspective of the horizon.

Chapters five and six deal with the qualitative differences between North, South, East, and West. The author recommends that each chapter be first read through to get a sense for the whole and then studied paragraph by paragraph.

Five star maps are included to illustrate Chapters eight and nine to help the reader during actual star observation.

Additionally the final two pages consist of a Planisphere and directions how you can use it.

In this booklet the starry heavens are described as they present themselves to an observer at about 45^0 Northern latitude.

David Mitchell
John Petering
Editors
May, 1994
Boulder, Colorado

The Study of the Stars - Part I

I. Introduction.

"Stars once spoke to human beings.
Their becoming silent is World-Destiny."
Mankind and the World of the Stars
Dornach 31 Dec. 1922 - GA 219

With these words Rudolf Steiner has described the whole tragedy of the modern person, who, unless they are an astronomer and have made the stars an object of special study, have lost the real knowledge of the Stars.

More and more the spatial distances between the stars have come into the foreground of a method of research which probes into the space of the Universe by means of ever bigger telescopes. Mechanical conceptions, derived from conditions here on Earth, are applied to the constituency and movement of the stars, but the concrete and living connection between the Earth with all her organisms and the Cosmos has faded away. This situation is not noticeably altered through the very considerable popular and semi-popular astronomical literature, or through articles in our daily papers describing the sky for this or that month. What has been presented in such publications is often admirable especially so in view of the fact that astronomy has risen to inconceivable heights of abstraction.

The "Studies" here presented have started from a very simple and practical question to which no real answer was found in any books available. Rudolf Steiner bases all his teaching about agriculture on the concept of the farm being a living individuality. This concept leads of necessity to that of a living Earth; a living Earth must be recognized to be a "Star among Stars." In our farm (and truly also in every member of it, in soil, plants, animals as well as in the human beings) we must discover the working-together, the presence or absence of a balance between earthly and heavenly, between terrestrial and celestial forces. It was felt imperative to develop a clear knowledge of the Cosmos, based on experience accessible to anyone irrespective of previous training etc. It must be clearly understood that these "Studies" do not contain any new astronomical discoveries. They differ however, fundamentally in their way of approach. All statements, pictures, exercises should be actually experienced by the reader. If they find it impossible to do so, the writer has not achieved what he set out to do, and his attention should be drawn to this fact. He will then take the earliest opportunity to clarify misunderstood and unclear passages.

That the Earth can be considered to be alive, many a farmer and gardener will admit, even if they have not yet thought of it before. But that the Stars should have something to do with the farm seems so strange that much greater efforts are needed to bring this concept to a living experience. But very much indeed will depend upon realizing that all life on Earth depends on the healthy equilibrium between the forces of Earth and Heaven. We owe it to the fact that we live in the 20th century that we should be conscious of what we do; and, only to the extent to which we strive for this consciousness in these matters can we claim to be modern human beings.

In looking around for help to answer this question, many books were consulted, the study of which only too often emphasized the fact that the stars have become silent. Yet, seeking nothing other than to know something about the stars and their message for us, no bridge was found between the human being and the surrounding Cosmos. One does meet such delightful books as Lawrence Edwards' "The Spangled Heavens" (Gerald Howe Ltd. London, 1933), but in reading it one becomes all the more aware of the abyss between the world of the Stars, and the human being here on Earth. Edwards describes in simple and beautiful pictures the Solar System with its Planets, the World of the Fixed Stars etc. What is missing, however, is the relationship of all this to the human being. As the case is presented, we seem to depend on telescope and mathematical calculations if we want to know something about the Universe. But, a single walk on a starlit night, and still more even the smallest efforts to understand the seasons, the phases of the Moon etc. can make us realise that there is also another way to the stars. Some friends at Clent Grove felt this very strongly -- and out of their co-operation, grew the plan to issue these "Studies."

Valuable help was derived from books, some of which are mentioned below. The help however, consisted more in the communication of facts than in the development of a way of approach. This latter is owed almost entirely to books and lectures by Rudolf Steiner. Their number is so great that it cannot be attempted to enumerate them. On request, detailed references and general recommendations will gladly be given. Of other helps, used with gratitude, the "Astronomical Letters" by Dr. Elisabeth Vreede, which are now unfortunately out of print, must be mentioned in the first place. Here an astronomer shows how Rudolf Steiner's Spiritual Science can bring again to life what has once been a noble art and has fallen into a highly specialised mathematical-physical science, inaccessible to the layman. The "Astronomical Calendars" by the same author contained in their German editions (now no longer published) most valuable articles by Dr. Vreede and others. Some have been translated into English. With special gratitude her chief assistant Joachim Schultz must be mentioned. (see also "Planisphere" below).

Dr. Hermann von Baravalle, formerly a teacher at the Waldorf School in Stuttgart, founded by Rudolf Steiner, has written (in German) an Introduction into the Phenomena of the Starry Heavens (Die Erscheinungen am Sternenhimmel, Dresden 1937). Much valuable guidance has been received from this book which is warmly recommended to those making a fuller study of astronomy. But even this book does not seem to build that bridge which the farmer and gardener wants. Of further books used, only two more should be mentioned which have been consulted a great deal: J. Plasmann, Das Himmelsbuch (Berlin 1925) and H. P. Hollis, Chats about Astronomy (London, I. Werner Laurie). [see also, Peter Lum, "The Stars in our Heavens" Pantheon, 1948 and Norman Davidson, "Astronomy and the Imagination," Routledge Kegan & Paul, Boston, 1985]

It is expected that the following two publications are in the possession of the readers of these "Studies", as frequent reference will be made to them:

(1) Movable Planisphere, in connection with the East, West, North, and South Horizon, by Joachim Schultz. (Mathematical-Astronomical Section, The Goetheanum, Dornach, Switzerland). [see adaption included in Appendix A]

(2) Sternkalendar (Star Calendar) annually, Easter to Easter, originally edited by Dr. Elizabeth Vreede, now edited by Suso Vetter, from the Mathematical-Astronomical Section, The Goetheanum, Dornach, Switzerland [through the Anthroposophic Press, Hudson, NY. USA, or the Rudolf Steiner Bookshop 35, Park Road, London.]

It should be quite clear that we are going to start in a very simple way with the observation of the sky, beginning with the Horizon, "where Earth and Heaven meet." For our immediate experience the Earth stands still and the starry heavens move around us. To base our study on this stationary Earth, appearing to us as a flat disc, does not mean a return to out-of-date conceptions, but it is felt that only in such a way can we study the stars free from theories and abstractions. As this Course progresses, we shall have no difficulty in introducing the concept of the Earth as a globe rotating round its own axis. Theories and abstractions are not bad in themselves, we are only faced with the undeniable fact that on account of them the ordinary human being has lost the language of the Stars. Let us try to listen to this language quite simply and modestly. There is no greater help towards realising both our individuality and our smallness in view of the majesty of Nature than by stepping out into a starlit night. According to laws unknown to us as yet, the Stars move around us. But it is the human being who looks at them, who can point to them. We never reach to them with our hands, but with our heart and with our thinking we can come close to them, so close that we recognise our origin and home in their world. The original question of how to experience the Earth as "Star among Stars," blends into the quest for understanding our origin and our being.

Having started from the relation between Earth and Cosmos, we recognise now:

COSMOS - HUMAN BEING - EARTH

as three components of one picture. In their body, the human being carries these three -- the human body is the key to an understanding of them.

The human head pictures in its spherical form the heavenly globe. The head is quite earthly in its bony material, but in its life it is far removed from the earth. Quietly we carry it through our life, the very image of the quiet movements of the stars. Our thinking, through which we can free ourselves from the world of the senses, has its physical support in the brain. At the other pole of the human body we have the ***metabolic-limb-system***. With our legs we place ourselves into the forces of earthly gravity. With every step we try to escape from, and yet fall back again to the Earth. In our metabolic processes we connect ourselves materially with the substances of the Earth. Cosmos and Earth thus actually live within us. As human being, we experience ourselves in that system which Rudolf Steiner (the discoverer of the Threefold Nature of the Human Being) calls ***the rhythmic system***. In the rhythmic circulation of the blood we have something quite un-earthly, overcoming the forces of gravity; in our breathing we communicate freely with the outer world. When the heart beat stops and the breath ceases, life has left the human being. Although being limbs, our arms have a closer connection with the Rhythmic System than with the Metabolic System. Unconscious of the forces of gravity, we can use our arms, can work with them on and with the material world, can meet our fellow human beings in salute or handshake, but we can also point to the greatest heights and furthest distances.

In these "Studies" we want to make full and practical use of the truth of Goethe's statement "Man by himself, insofar as he makes use of his healthy senses, is the greatest and most exact physical apparatus which exists." The vertical position of our body will be the chief direction to which we shall relate all directions on Earth and among the Stars, as in the vertical position our body is the most perfect expression of the balance between Cosmos and Earth. Our senses, notably that of sight, will bring us into communication with the Stars and with the Earth on which we stand. We shall use our arms to point to the Stars, to follow their movements, to measure angles etc. All the time we must never forget that it is the human being who looks at Earth and Stars. The question "Is the Earth a Star among Stars?" is asked by the human being. The answer cannot be found but through the human being.

II. The Horizon and Other Great Circles

Introduction

We begin our Study of the Stars by trying to take our bearings when we stand under the starlit sky. We do not know yet where North and South, where East and West are. ***The only certainty we have is that we stand on a point on Earth.*** We look up to the sky and see a multitude of stars, some brighter, some less bright. Our first experience is that they shine down to us from a majestic dome extending around us. We see stars to our left and right, in front of us and, as we can ascertain when turning round, behind us as well as above us. We do not know how far they are away from us, but all stars seem to have the same distance, and we stand in the centre of this celestial globe.

Earth and Heavens meet in the Horizon (from the Greek word meaning: to form a boundary), and the study of this Horizon has to be our first task.

Horizons

The study of this Course will be made easier if we select for our observations one particular place, easily accessible from our house, and if we make ourselves familiar with the outlines of its Horizon. An Ideal Horizon we shall only have on the sea or in flat and open country. In many cases hills, high buildings, trees etc. will tower high up and will greatly restrict our range of vision. We must realise however, that under whatever conditions and in whatever locality we may look at the Horizon, however irregular it may be in its outlines, it is always in this Horizon that Earth and Heavens meet. By changing our standpoint we can often greatly alter our Horizon, as we can easily find out by going, for example, on the roof of a house and comparing that Horizon with the one we have when standing on ground level. We can never see below the Horizon we have at any given moment.

In the following Studies we shall assume that we stand in a plain, or are on the sea. In this way we can base our descriptions on an Ideal Horizon and can simplify our drawings. But it is nevertheless necessary to become familiar with the Horizon of the place one chooses for regular observations, and to remember characteristic trees or other special marks. It is a very good exercise to draw the outlines of the Horizon. In doing so, many valuable observations can be made, and some difficulties which might otherwise arise in studying this chapter, can be avoided if we have a good picture of our Horizon and compare that with the Ideal Horizon used in these our Studies.

It will be essential that we always try to observe the phenomena we study, under the conditions of our own locality. It is a good preparatory exercise in observing, to sketch the Horizon. One of the next chapters will deal with the movements of the Stars near the Horizon and we shall base the discussion of these phenomena on the Movable Planisphere referred to in the Introduction. It will be a great help if the reader has accustomed himself to using this Planisphere in conjunction with their own Horizon.

The Ideal Horizon (marked in the Planisphere as a fine silver line) seems to be everywhere equally distant from us, or in other words: we experience ourselves in the centre of that surface of the earth which is limited by our Horizon. For the rest of this chapter the Planisphere is not of immediate use, as it has been constructed in a way which we shall study in more detail only much later. The following discussion **needs to be experienced in actuality** and in connection with the Horizon we see in Nature. ***The careful study of the qualities of the Horizon is most essential because it is just there that our earthly experience ends.*** During our observations, we need not concern ourselves with what lies between us and the Horizon, as the horizon remains relatively constant, unless altered by human beings (new buildings, etc.) We can take the Horizon as something given, as something unchanging. Earth and Heavens meet in the Horizon, but we do not yet know at this point of our Studies, where on the Horizon lies the South, where North, where East and West.

Let us stretch out our hand towards any point of the Horizon and then follow the Horizon line round towards the right or left. If we have made one complete revolution, we return to the point of departure. Thus the Horizon reveals itself again as a circle. The first time, we are able to experience the circle-nature of the Horizon from the elementary fact that we stand in the centre, at equal distance from all points of the Horizon. But how does this circle actually appear to our sight? We do not see the Horizon as a circle but as a straight line. We cannot notice any curvature whatsoever. Some readers might be inclined to think of the Horizon bending down on their right or left. This is an illusion due to the fact that we are apt to imagine a globe as seen from outside. Here however, we stand on the surface of the Earth and the Horizon is around us. Should there be any doubt, we can take a straight ruler and hold it with outstretched arms parallel to the Horizon, and we shall always be able to cover the Horizon line quite exactly with it, assuming of course that we stand in a plain or are on the sea.

Great Circles

Thus the Horizon is a circle which appears to us as a straight line. This is the case with any circle in the centre of which we stand, or better, in the centre of which our eyes are. We shall call such a circle a ***Great Circle***. Later on we shall meet many more such Great Circles. We can always recognise them by either finding ourselves the centre, or by seeing them as straight lines. It is most essential that we have a real experience of what a Great Circle is, how we see it, and how we can describe it with our arms. If we hold a hoop in such a way that our eyes are at the centre of it, we shall easily recognise the nature of a Great Circle. At this point special emphasis is laid on the help afforded by the use of our arms. When in the following pages such phrases are used as "We point to this or that star" or "We describe with our arms the path of this or that star" etc., the reader is asked to do this quite concretely. In the Introduction (above, page 3) Goethe was quoted as having said that the human body is the finest instrument we have. We must now learn the use of this instrument.

Horizon Exercises

Let us stand upright and stretch our two arms in opposite directions so that the line joining our two hands always passes through our eyes. Now let the arms move so that the hand which comes within our vision appears to describe a straight line. So long as the above conditions are not altered, this movement will inevitable describe a Great Circle. Such a position is approximately indicated in Fig. 1.

Figure 1 - Describing a Great Circle with the arms.

It will easily be seen that an infinite number of such Great Circles is possible, but that one Circle among all others occupies a unique position, namely that *it coincides exactly with our Horizon.* This special position is shown in Fig. 2.

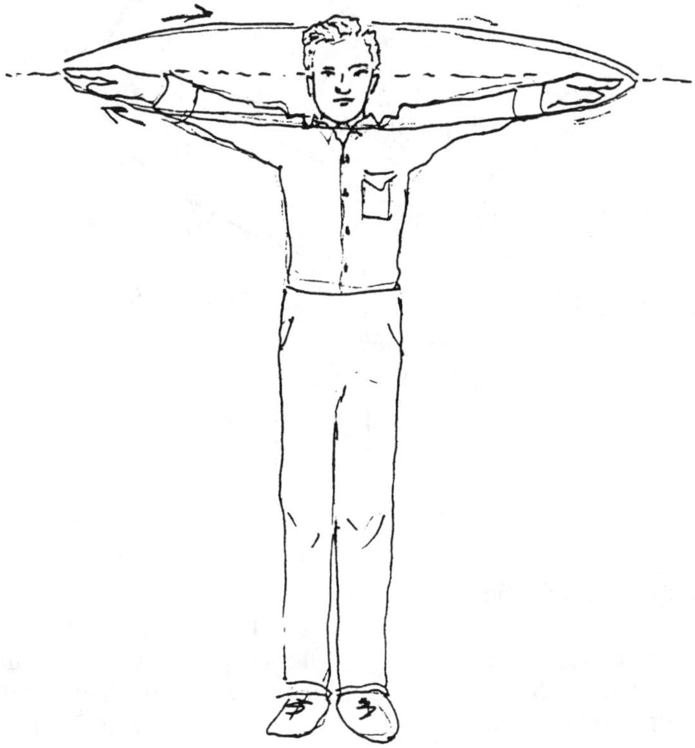

Figure 2 - Horizon Circle forms a line, just at eye level.

THE HORIZON AS A GREAT CIRCLE

The difference between the Horizon and all other Great Circles is this: in order to see all parts of the Horizon we will need to move (turn), but all the time our outstretched arms, pointing to the Horizon, do not change their angle with the vertical direction of our body, i.e. a Right Angle. We can see all parts of the Horizon by looking straight forward, we merely turn around our own vertical axis. We must really do this once and experience that all we have to do in order to see the Horizon in its totality is to make one complete turn around ourselves, all the time looking straight ahead of us.

OTHER GREAT CIRCLES (STAR PATHS)

All other Great Circles are cut in two parts by the Horizon, one half is above and *visible*, one half below and *invisible*. There are two possible ways that the other half can come into our vision.

One is that a Great Circle would ***rotate along its periphery*** (clockwise or anti-clockwise). Let us assume the Great Circle in question would turn in the direction of the arrows (left to right) as shown in Fig. 3. After a first quarter turn, a point on the periphery of the Great Circle which earlier crossed the Horizon on our left (A), is now somewhere in front of us (B). What was in front of us, now approaches the Horizon on our right, and is on the verge of disappearing. Then, we make another quarter turn in the same direction and bring the point in front of us to the Horizon on our right (C). In other words: in two quarter turns (or one half turn) or by rotating the Great Circle through 180° along its periphery, the point previously near the Horizon on our left has changed to the Horizon on our right. With the words 'A star moving along a Great Circle' we mean such a movement along the periphery.

If we now follow this movement with our eyes we soon realise that we have to continually alter the height (or altitude) of our line of vision; our movement is totally different in quality from that when we simply surveyed the level Horizon.

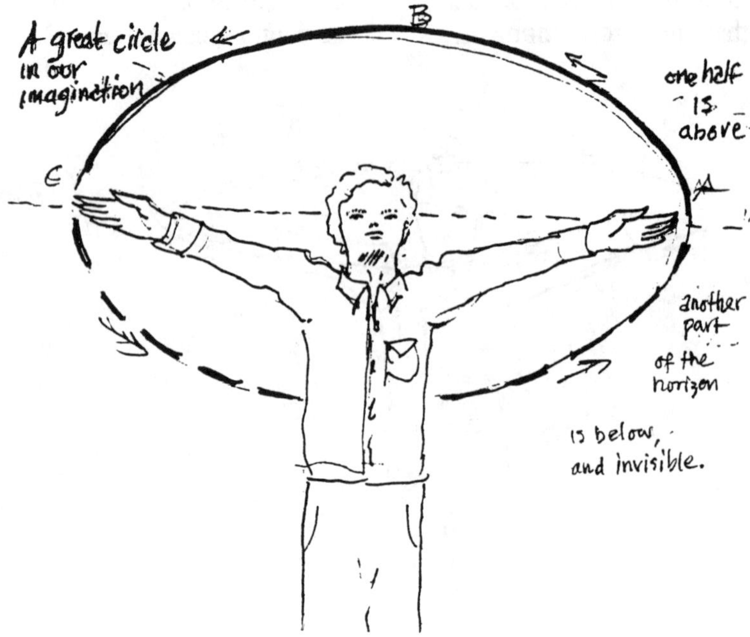

Figure 3 - Visibility and invisibility on a Great Circle

But there is *a second possible way that points on that half of the Great Circle which is below the Horizon can come into our range of vision.* Let us **take a hoop** to represent a Great Circle, and hold it in an inclined position, corresponding to position A (in Fig. 4, below).

Holding the hoop in our finger tips, and taking care not to alter the position of our arms or hands, we can -- by a simple finger movements only -- make the hoop tip either up (B) or down (C) into positions B and C of Fig. 4.

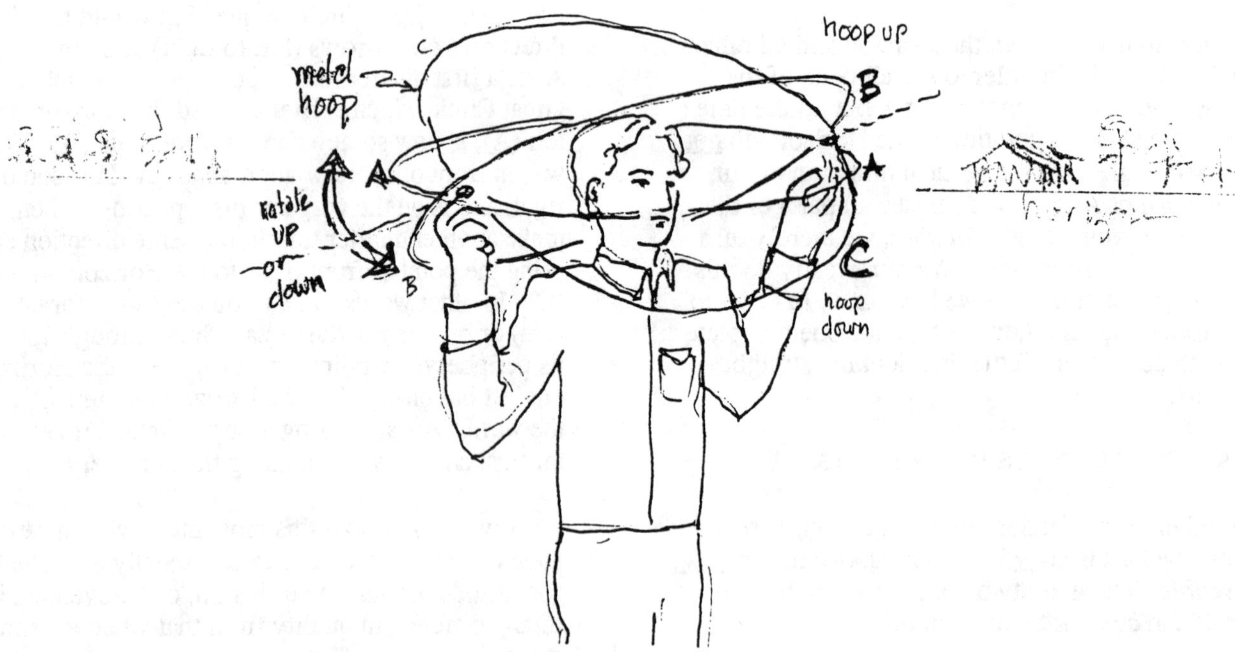

Figure 4 - Tilting a hoop to study various Great Circles

Thus, the Great Circle forms an angle with the Horizon which may be anything from a fraction of a degree to a Right angle. The points where Great Circle and Horizon cross one another are always diametrically opposite each other, namely at our finger tips. We can now alter this angle between Great Circle and Horizon, without turning the Great Circle along its periphery. Now imagine the hoop moving (down) from position B towards position C. There comes one moment when Great Circle and Horizon circles coincide (position A), when the angle they form with each other is 0°. If we continue to move the hoop down in the same direction, we shall find that the moment we go past position A, that half of the Great Circle, which was below the Horizon and behind us, is now above it; and, what was above, is now below. But a further very important change has taken place too: If we had at first the visible part of the Great Circle <u>in front</u> of us, now the part of the Great Circle above the Horizon lies <u>behind</u> us.

In the case of a ***rotation along the periphery***, new portions of the Great Circle were continually curving up above the Horizon and into our range of vision. In this second case an instantaneous change between visibility and invisibility, between being above and below the Horizon takes place. The difference in the dynamics of these two movements as experienced by us, is of the greatest significance and should be livingly grasped by the reader before going any further.

All this sounds abstract and may appear at first very difficult to understand. But if one tries to do these exercises, one will soon discover what is meant. In going through these exercises we get familiar with the Horizon and with the concept of a Great Circle. What is more important still, ***we begin to use our body and our experience as an instrument of observation***. It is therefore hoped that the readers of this Course will not jump over this section. The writer is only too well aware of the fact that to show these exercises is much easier than to describe them in words. It will perhaps help us to see the necessity of a very close study of the Horizon if we realise that in the Horizon, the Earth and Heavens meet. We shall feel our way, at it were, from the Earth into the sky. Let us make as clear a picture as possible of what the Earth looks like where it meets the sky, the World of the Stars. High up in the realm of the Stars we shall have no longer an Horizon to which we can relate the celestial bodies in their position and movement. Already in the next chapter we will notice how lost we can feel in the midst of the stars, therefore let us become aware of the special characteristics of the Horizon and clearly establish our relationship to it.

III. The Zenith and Nadir

We have recognized the Horizon as that Great Circle in which Earth and Heavens meet. We can point towards it without outer aid, as *we carry in our body the measure of a Right Angle with the vertical direction.* Let us fully experience the fact that we can find the vertical direction with the help of our own body, not requiring any other instrument. We shall find it easier to stand "vertically" if we raise our heels slightly off the ground. If we do this our spine points towards that point in the sky which is directly, vertically above us. This point is called the ***Zenith*** (an Arabian word meaning "the apex").

ZENITH

Let us make this exercise: In order to find a point directly, vertically above a certain point on the ground, outside us, we need an instrument, even if it is only a simple plumb-line. The moment we stand on this point, we no longer need a plumb-line, but can determine through our own body the direction, in which all points vertically above the point on the ground are to be found. Such little exercises help us to gain confidence in our body as this finest instrument of which Goethe speaks.

There is a great and fundamental qualitative difference between our connection with Horizon and Zenith. The Zenith is the point in the sky vertically above us. *We cannot see this point* in the normal position of standing upright. If we bend our head so that we look towards our Zenith we can make two important observations. 1) We are not so absolutely sure exactly where the point directly above us lies - unless we lie flat on our back and look straight upwards. 2) If we bend our head backwards for any length of time, we can notice a slight (or even big) change in our consciousness. Some people even get giddy very quickly. Furthermore we find it much more difficult to locate the direction from which a sound meets our ear. On the other hand, when trying to connect ourselves with the Horizon we <u>are</u> fully conscious of our upright, truly human position -- we can look towards the Horizon with ease.

We can also describe this difference in another way: We are intensely conscious of the Zenith, the more so the more quietly we stand upright. We become aware of the Horizon through the activity of looking and of stretching our arms. We must only observe how difficult it is to point with our arms to our Zenith. But in trying to become conscious of the Zenith, we shall have the great satisfaction of realising a sense of balance, or equilibrium. In the coming chapters, the more we follow the movements of the stars, the more important it will be for us to maintain this sense of balance.

NADIR

Directly, vertically below us lies a point equally well defined as the Zenith. This is called the ***Nadir*** (also and Arabian word, meaning "Foot-point"). What has been said about the Zenith and our connection with it, applies to the Nadir as well. We shall see presently that the realisation of the Nadir is very useful.

Perhaps it is here appropriate to draw attention to the fact that, in these "Studies," we are not concerned with distances, neither on Earth, nor in the sky. We look towards the Horizon, and may or may not know how far it is away from us. Actually, the range of our vision, or in other words the distance of the Horizon, can change rapidly with our position. But this need not concern us here. Nevertheless, we can always realise ourselves in the centre of the Great Circle which we call the horizon.

Similarly, it is not necessary to express in figures the distance to the stars in the sky. It fully suffices for our observations and experience that we are able to point towards a star. Our vision establishes a definite and clear connection with each star we pick out. In this way we feel also clearly connected with the Zenith above us and the (invisible) Nadir below us. [We deal with directions, not with distances.]

VERTICAL GREAT CIRCLES

Now, we lift our arms so that our hands meet above our head. Then, with outstretched arms we describe a vertical Great Circle, i.e. we (or strictly speaking our eyes) are the centre of the circle described. We can form an infinite number of such vertical Great Circles, each time however, we face another point on the Horizon. Every Great Circle which goes through the Zenith passes also through the Nadir and all these vertical Great Circles going through Zenith and Nadir have one characteristic in common: <u>they cross the Horizon at a Right Angle</u>. The Horizon naturally halves all these Great Circles, but it is important to realise that they all cross the Horizon at a Right Angle, or in other words, that they stand vertically with respect to the Horizon, just as we stand vertically on the ground.

It is necessary that we have an immediate experience of this. Quite intentionally we leave out of account geometrical proofs which could be furnished for all these statements. Much will depend on whether we can realise these facts both through being conscious of our body and through our sense of sight and imagination.

Just as every observer has his own Horizon, so also every human being has his own Zenith and Nadir.

If we now watch the celestial Zenith for even a comparatively short period of time, or if we look up to the Zenith at intervals during one evening, we become immediately aware of the fact that ever different stars stand in the Zenith. From this observation we learn that in spite of the fact that we can so clearly define the Zenith through our body, zenith and horizon alone do not provide an adequate orientation in the sky among the stars. Something further is needed if we want to come to such an orientation. With this we shall concern ourselves in the next chapters.

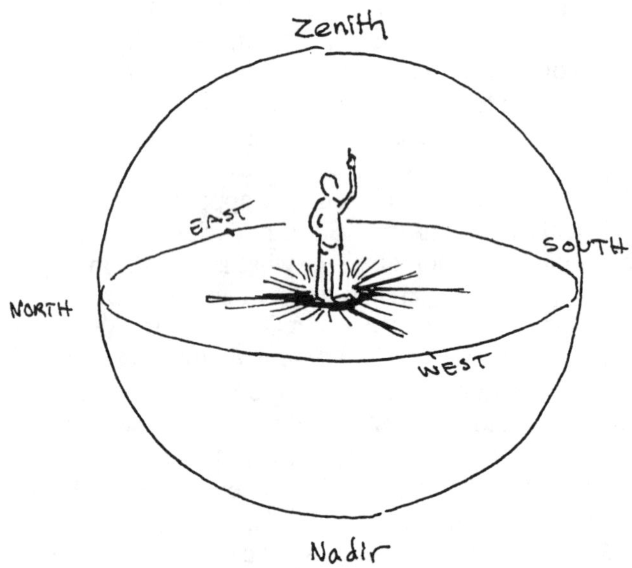

Figure 5: Zenith and Nadir about us

The Study of the Stars - Part II

IV. General Remarks on the use of the Movable Planisphere.

The following chapters will be based on the observation of the stars in their movements. We are not always able to watch the stars themselves however, and in such cases the "Movable Planisphere" by Joachim Schultz, in connection with the East - West - North - South Horizon will prove very useful. [See adaption of this provided in Appendix A.]

It is a peculiarity (anomaly) of this Planisphere that it shows a specific, actual Horizon. This is particularly valuable to us as we shall begin our study of the stars *from* the horizon. It will here suffice to explain the use of the Planisphere for the first, more qualitative observations. In Fig. 6a, the planisphere shows us a certain part of the horizon. That this southern horizon is an irregular line is due to the fact that the actual, visible horizon about the Goetheanum in Dornach, Switzerland has been chosen. The high elevation between "Osten" and "S.O." is the Gempen; the little building in the left-hand corner, the ruins of Castle Birseck. The "ideal" horizon of the previous chapters is the thin, silver oval line within these irregular arcs.

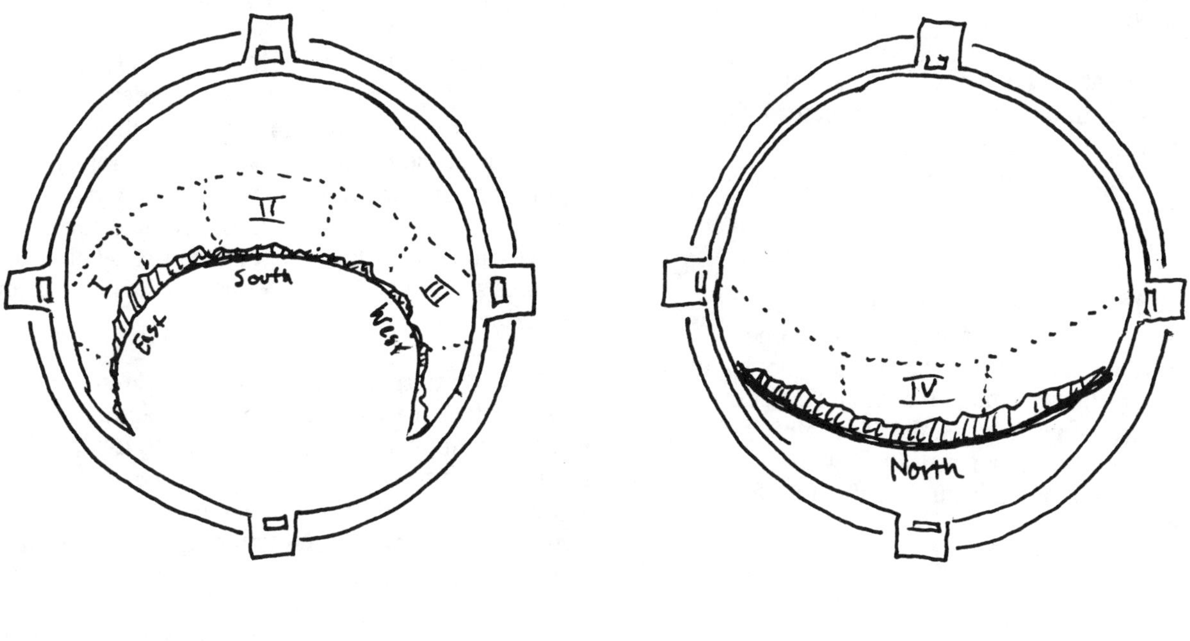

Figure 6 - Planisphere by Joachim Schultz

We shall not here explain in detail why the (true) horizon line has been drawn in this unusual curve. Suffice it to say that the horizon extends around us, the heavenly dome around and above us, and any attempt to reduce this three-dimensional space into a two-dimensional plane must lead to compromise -- somewhere the picture must be distorted. When, later on, we make fuller use of the Planisphere we shall show how to bring back this two-dimensional picture into the reality of three-dimensional space.

When using the Planisphere for the study of the star movements near the horizon, we must limit our observation to the lower regions of the heavens, which have been indicated by the dotted line running roughly parallel to the horizon. The three sectors I, II, III along the southern horizon (illustrated in Fig. 5) show the approximate regions of the sky, and the portions of the horizon to which we will restrict our observation. These sectors I, II, III correspond to Figs. 7, 8, 9, in Chapter V.

All we have to do now is to turn the movable part of the Planisphere (the round star map enclosed in the horizon template) in the direction of the arrows, i.e. clockwise. We watch how the stars move in sectors I, II and III respectively. The details of these movements will be discussed in Chapter V. and will be recognized as corresponding to the star motions studied in Figs. 7, 8, and 9 respectively.

If we now look at the back of the Planisphere (Fig. 6.), we must limit our observation to sector IV (Northern horizon). When the movable part of the Planisphere is turned in the direction of the arrows (counter-clockwise), we can observe what will be shown in Fig. 10, Chapter V.

Later on, full instructions will be given for a much more extensive use of the Planisphere. It would be premature to explain here already the constellations and the meaning of the figures around the circumference. At the present stage of our studies, the Movable Planisphere is only used to demonstrate how the stars move near the horizon, and it is only a substitute for observations out of doors.

V. The Movement of the Stars near the Horizon.

When we now begin with the actual observation of the stars we must realise that we do not yet know where North, South, East and West are. This we have still to learn. As was suggested in Chapter IV, we shall limit our first observations to the lower regions of the sky. It will be quite important to begin the study of the heavenly phenomena there, where heaven and earth meet. We shall thus be able to link the celestial phenomena with our experiences here on earth.

Note that what is described in here Chapt. V need not necessarily begin with Fig. 7, followed by Figs. 8, 9 or 10. It could equally well be studied beginning with Figs. 8, 9, or 10 so long as that sequence is followed which will be outlined hereafter.

When we observe the movements of the stars ***near the Eastern horizon***, pictured (in Fig. 7, opposite), we shall notice that here ***all stars move upward in a slanting direction from left to right***. In order to see this, one can either watch the stars continuously during the period of an hour or two, or one can look into this region at regular intervals of, let us say, a quarter of an hour. The chief impression is that all stars we see move upwards, rise. Secondly it is important to notice that they rise at an angle, not vertically.

If we now turn to the right by 90 degrees (i.e. a Right Angle) and face South, we face that part of the southern horizon pictured in Fig. 8. A different view presents itself here. We see stars both rising and setting. Other stars move right across the picture somewhat parallel to the horizon without touching it, and we can notice that the arc of their movement becomes shallower the greater and the distance from the horizon.

Now we turn once more by a right angle to the right and face the western horizon as pictured in Fig. 9. In this position we have our back to the East, to Fig. 7, and our left arm, when stretched out, points South, towards Fig. 8. In the West, we see stars only setting, or moving downwards and to the right. The picture is an opposite one to Fig. 7.

In conclusion (Fig. 10), we face the northern part of the horizon, opposite Fig. 8. Our left arm points now towards Fig. 9, where stars only set, our right arm towards Fig. 7, where stars only rise. Here a picture confronts us in which stars set and rise. In the left half of this picture stars approach the horizon, while they rise in the right half.

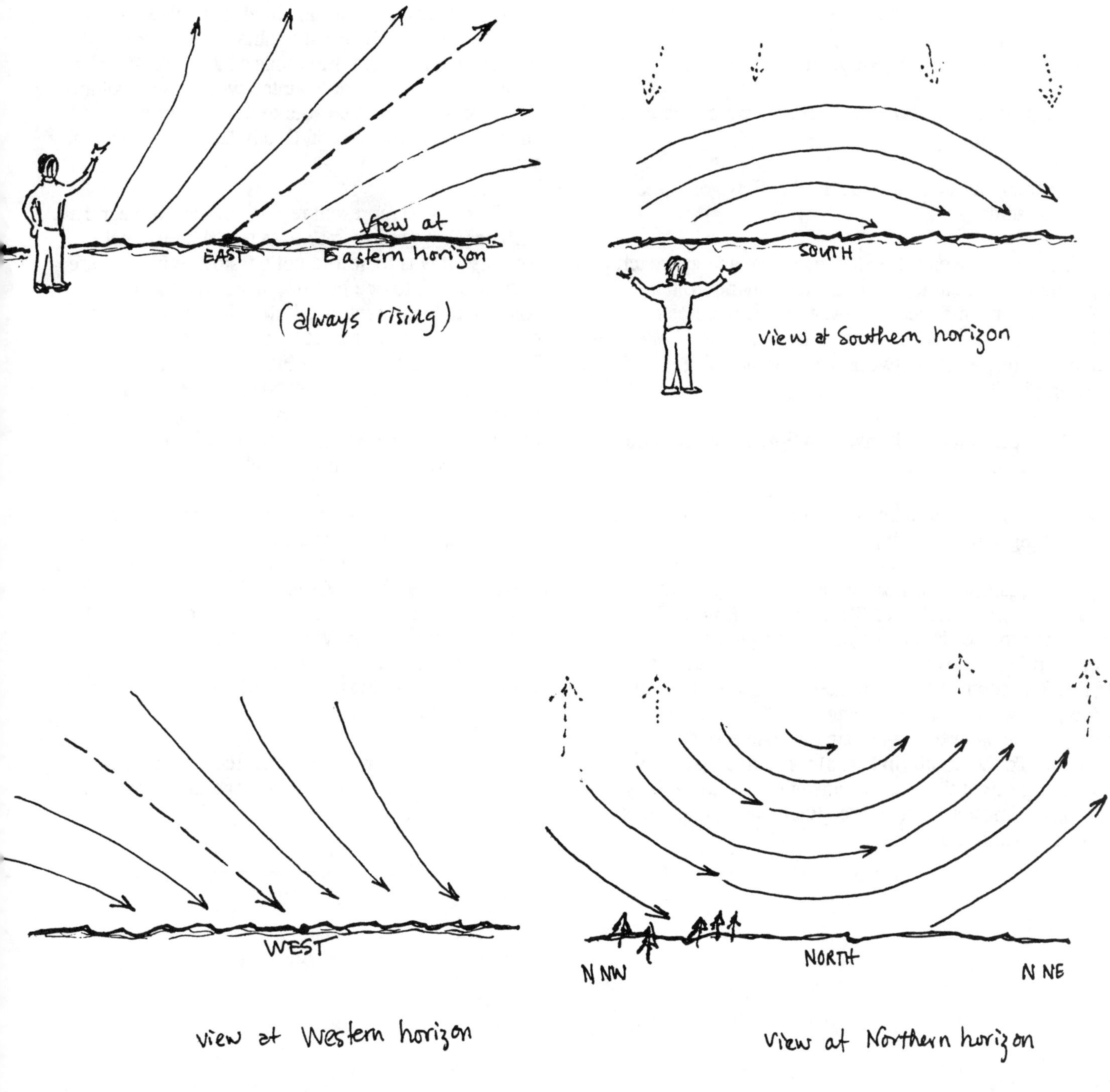

Figures 7 - 10: Star motions near the four horizons

In the centre stars can be seen approaching the horizon and leaving it again without touching it. The further the stars in the centre of the picture are up from the northern horizon, the more curved is the arc of their movement, (open towards the upper part of the sky).

We have faced the horizon in four different directions and we must make every effort to experience the contrasts between them.

We summarise these four pictures as follows:

- Near the Eastern horizon, in Fig. 7, we only see stars rising.

- Near the Western horizon, in Fig. 9, we only see stars setting.

The angle at which the stars rise from and approach the horizon is the same, but it is important not to take the fundamental difference between these pictures too lightly. We should consciously experience the qualitative difference between the regions in Fig. 7 and those in Fig. 9.

- Near the Southern horizon, in Fig. 8, we see stars rising and setting.

- Near the Northern horizon, in Fig. 10, we see stars setting and then rising.

Here again we must try to follow these movements as consciously as possible. To the South, (in Fig. 8) stars leave the horizon and return to it again; to the North (in Fig. 10) stars approach the horizon and rise again. The contrast between these pictures is felt more strongly if we watch the movements of the stars in the centre of the picture, at a certain distance above the horizon. To the South (in Fig. 8) we notice that with increasing height the arcs become ever shallower, they approach almost a straight line. Each one of these arcs is open towards the horizon.

To the North (in Fig. 10), however, the arcs become more curved with increasing height and they are open upwards towards the higher regions of the sky.

The star movements to the South and to the North (in Figs. 8 and 10) are differently orientated: to the South, they tend towards the earth, to the North, towards the sky.

If we study the star movements in this way we become aware of the great qualitative contrasts between the different directions of the sky. When we have made ourselves acquainted with these qualitative differences, we shall be able to develop a living knowledge of what North, South, East and West really are.

Figs. 7 - 10 represent regions of the sky near the horizon, as we see them, i.e. as two-dimensional pictures. It was therefore not possible to bring these four aspects into one picture, just as little as it is possible to look towards all directions of the sky at once. We must not forget that we proceeded from Fig. 7 to Fig. 8 to Fig. 9 to Fig. 10 by turning in each case around our own axis by 90 degrees. This movement of ourselves as observers must not be ignored. [The teacher could paste the four patterns on the corresponding walls of the classroom - Ed.]

We will have realised by now that to restrict our study of the star movements to the regions near the horizon only, leaves us dissatisfied, but the attempt to make a picture of the whole of the sky with ourselves as observers in the centre, results in a complicated drawing. We purposely proceed slowly so that proper foundations are laid, for an understanding of more complicated drawings, the use of which will later on, be unavoidable.

Our next task will be to link these four horizon pictures together. This will be done in Chapter VI, which follows.

VI. The Movement of the Stars over the Whole Sky.

For a proper understanding of this chapter, much will depend on being aware of the fact that we stand in the centre of the celestial dome, and that we stand in the centre of the horizon. We must develop a feeling for the space around us. By standing upright, we connect ourselves with the zenith, by looking straight in front of us we connect ourselves with the horizon. It will be important that we retain the consciousness of this space, which can be called a stationary one.

In the last chapter, we saw as a picture before us, how the stars move near the horizon, and we were able to analyse these movements into four distinct regions. If we now want to *bring them together again*, to synthesize them, we can no longer remain the onlookers in front of a picture but we must realise the star movements ***through space***. We can do this best by pointing with our hands towards the stars and following their course with the movement of our arms. We also need our eyes and we must become aware of the direction in which we look.

If we stand so that we face the southern horizon (as in Fig. 8), then our outstretched *left* arm points to the East (Fig. 7), our outstretched *right* arm to West (Fig. 9). (The expression "Fig. 7" etc. must here and in the following be understood as implying "the region of the sky pictured in Fig. 7" etc.)

Without altering the position of our body, with our left arm we trace the movement of the stars, from East to South, higher and higher into the sky, until the left hand reaches a point somewhere in front of us at a certain height above the southern horizon. Then, we bring our right arm up into the same position, and move *it* down towards the western horizon, so that it enters the region of Fig. 9.

At first, this movement from East over the South, to West (Fig. 7 to Fig. 9) should be done quite qualitatively, without following any particular star in its course. A feeling must be developed for the fact that the stars, rise on our left, reach the highest point of their course somewhere in front of us, and set on our right. It should be noted that the stars rising on our left do not rise all the way up to our zenith, but only somewhere in the region between the zenith and the highest star-courses in Fig. 8.

Having become familiar with this movement in general, we can now make more exact observations. First we make sure that we are facing exactly South (the centre of Fig. 8), and that our arms point exactly at a right angle to our line of vision. If we now move our left arm upwards at the angle at which the stars rise there, we can observe that we **describe a Great Circle**, i.e. a circle in the centre of which we stand, and the periphery of which we see as a straight line (See Chapter II, above). With our left arm we describe one quarter of this Great Circle, and continue downwards with our right arm through the second quarter.

The full significance of the uniqueness of this star-course becomes clear if we now describe the course of any other star. For example we chose a star rising to the right of the one dealt with above. In Fig. 7 such a star-course is marked with a dotted line. We are particularly careful to see that the angles are correct which this course makes with the eastern horizon on our left (Fig. 7) and western horizon our right (Fig. 9).

Such a course will no longer be seen as a straight line, it is not a Great Circle but it is clearly seen as a smaller arc. If we now follow the course of a star rising still further to the right from the original point on the horizon (Fig. 7), the smaller the arc becomes, and the more obvious it is that this is no longer a Great Circle. Very soon we find ourselves describing arcs belonging already to Fig. 8.

We can put this experience into different words. If we try to link up in space the movement in Figs. 7 and 9, we describe a Great Circle. This Great Circle crosses the horizon at Figs. 7 and 9, at a slanting angle. (If this angle were a right angle this Great Circle would go through the zenith, but only then.) All arcs extending from points South-east (to the right of the centre of Fig. 7), to points South-west (to the left of the centre of Fig. 9), are **smaller than semi-circles** (less than half a circle); they are also all **more strongly curved**. The more we move with these arcs away from the centres of Figs. 7 and 9 respectively, the nearer we come to Fig 8 and gradually we describe with our arms movements of stars which, in Chapter V, we learned to know as belonging to the southern region of Fig. 8.

Let use watch the position and experience of our instrument, the body. The movements of our arms, so far described, are easily made and fill us with a feeling of satisfaction. What we have done is to describe a Great Circle from our left to our right, the angle of which was determined by the movement of the stars in relation to the horizon. We have then described smaller and smaller arcs, and in doing so, we noticed that the highest point of each successive arc was nearer to the horizon.

Now we try to follow the course of a star which rises further North of East (to the left of that Great Circle linking Figs. 7 and 9). This arc rises higher than the Great Circle, and it sets to the right of the centre of Fig. 9. We continue in this movement (still always facing South, Fig. 8) and make the unpleasant discovery that our body becomes less and less able to serve as a willing and flexible instrument. We are forced to bend our arms further and further backwards, and at the same time we have to look higher and higher if we want to focus the highest point of each arc with our eyes. We furthermore observe that the arcs themselves become greater and greater.

We soon reach a point where we can no longer describe these movements. We now turn completely around and face North (Fig. 10). We have to do so very consciously, as so far we have remained stationary all the time facing Fig. 8. Henceforth, we completely alter our position and face the diametrically opposite region of sky and horizon. Our right arm now points where formerly the left arm pointed, and vice versa.

Supposing that the last arc we were able to describe before we turned about, rose from a point on the northeast horizon, half way between the centres of Figs. 7 and 10, and set half way between the centres of Figs. 9 and 10, then our left arm pointed to the rising, our right arm to the setting of this star. Now, when facing Fig. 10, we follow with our right arm the rising of the stars, with our *left* arm the *setting*. We move on the horizon to the left from Fig. 7, to the right from Fig. 9, and soon we discover ourselves describing star movements we saw pictured in Fig. 10.

The stars which rise in front of us curve upwards to the right, reach their highest position in the neighbourhood of our zenith, and moving downwards on our left setting very close to their point of rising.

Next we find that there are stars which only approach the horizon, but before reaching it continue their course in an upward direction. These circles become smaller and smaller until we find that there is one star which to the naked eye does not move, which is practically stationary.

When we have reached this point, we have described the course of all the stars (Figure 11, opposite).

Before we can make full use of these observations it is essential to establish a living knowledge of these movements. Two dangers must be avoided: on the one hand the reader might feel that he will never understand these abstract, complicated movements. But, he will do so, if he himself becomes active and conscientiously repeats these movements with his arms. It is fully realised that this chapter is by no means easy to understand at a first reading. It should be taken, rather as an explanation of exercises. One of the chief obstacles for an understanding is that the turning about from facing South (Fig. 8) to facing North (Fig. 10) is forgotten. We easily pay too much attention to the movement of our arms and the course of the stars and ignore that we as observers, have moved too. It is left to the initiative of the reader to try these exercises while lying on the back. Certain points are then more easily understood because we need not then change our position. (The feet would all the time point in the direction of Fig. 8.) A full description however of the exercises worked out for this lying-down position is not given, because some aspects are less obvious to us while lying down then while standing upright.

The other danger is a very modern one: we are so used to intellectual abstraction that this chapter may only be read and not enacted. One can of course grasp these movements merely intellectually. But the reader can be assured that he will miss a great experience by foregoing the joy of linking together with his arms and eyes what he first splits asunder.

We began with finding our connection with horizon and zenith. We were then forced to realise that the stars move in a way not easily followed in the orientation gained by consciously placing ourselves into the direction of horizon and zenith. We therefore divided the sky into one region near the horizon and the dome above it. Next we noticed that once we divided what is a whole in nature -- the starry sky -- we had to split it up further. We cannot make a picture of the horizon as a whole and arrived at four separate pictures. They have in common, that we as observers face them all, and that they are equal in value in spite of all their contrasts.

Purposely no attempt has been made to illustrate this chapter. At first the reader might be disappointed. But it must be realised where the difficulty in illustrating this chapter lies. The active human being must link the four pictures of Chapter V together, and he must do so in space, he cannot do it merely in a two-dimensional or mental picture. Such a picture would have to contain both the whole heavenly space and the human being. One can of course make such drawings, and later on we shall introduce this way of presenting celestial phenomena, but to do so already here would run contrary to the whole idea of this course.

Star Motions: Conclusion

In conclusion let us review once more what we have learnt so far about the movements of the stars.

- The whole of the starry heavens move around us.
- We have found one star course which is a Great Circle, extending between two points on the horizon diametrically opposite to each other.
- This Great Circle stands at an angle to the Horizon.
- All other stars move along courses parallel to this Great Circle.
- The points on the horizon between which the arcs extend, which are smaller than the Great Circle, move closer and closer together, towards the South.
- The points on the horizon between which the arcs extend which are greater then the Great Circle, also approach each other, towards the North.
- Eventually an arc is reached which touches the horizon in one point only. Above this, the courses of the stars are visible as complete circles.
- In the centre of these circles is one star which does not seem to move at all.

In Chapter VII (Part 3), we shall utilise the knowledge obtained so far to come to a definite orientation both on the horizon and among the constellations in the sky.

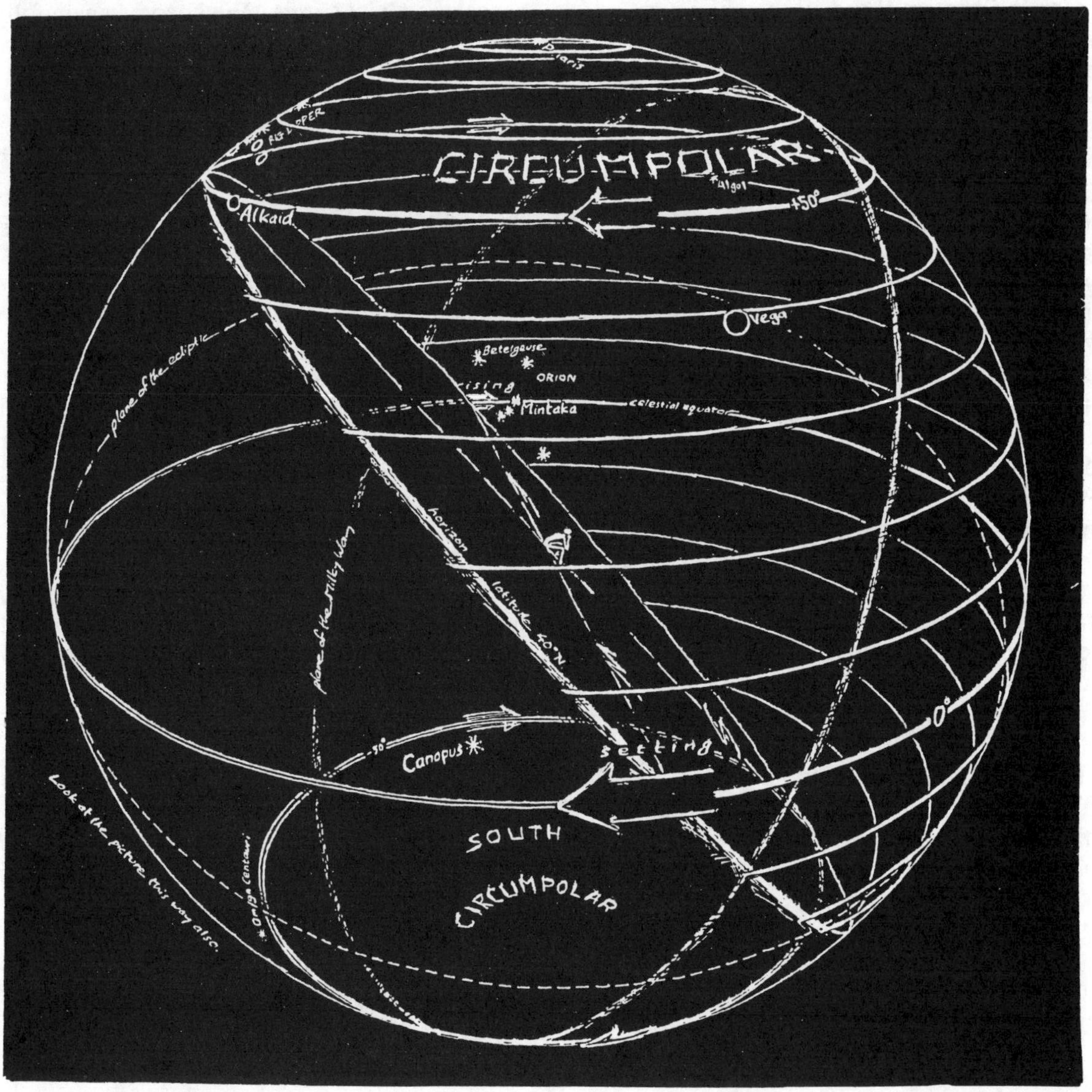

Figure 11

The Study of the Stars - Part III

VII. North, South, East and West; Meridian and Equator.

As upright human beings we find Zenith and Horizon and are able to link ourselves with them irrespective of locality, time, etc. Although Zenith and Horizon are thus clearly defined, we cannot face a specific direction on the Horizon or in the regions between Horizon and Zenith unless we drew upon our knowledge of the star movements.

We have seen in Chapters V and VI that the stars move obliquely to the horizon, with the exception of one star which does not seem to move at all.

Pole Star

We can find this star when we look up into the sky when facing the North (see Fig. 10, again). Let us now point to this star with our outstretched arm and thus draw an imaginary line between us and this star. This line (of which our outstretched arm forms a part) is an **axis** around which moves the whole of the starry heavens. This axis points towards **the Pole** of the starry heavens, the word Pole being derived from the Greek word for rotation. To be quite correct it must be stated that, to the naked eye, this star (called the ***Pole Star***, ***Stella Polaris***) *seems* to be stationary, whereas actually it circles close around the true Pole at a distance of about two Moon-diameters. This distance changes in the course of centuries quite considerably, but for the practical discussions of our present studies we are allowed to ignore this, as long as we remember this tiny inaccuracy.

The fact that the axis between us and the Pole Star is relatively fixed is of importance both for our orientation in the sky and on earth. Let us now connect ourselves with the Pole Star and with the Zenith by describing a Great Circle vertically through both these points.

Through any point in the sky we can place an indefinite number of Great Circles (see Chapter II, page 4). We should actually practise this again with regard to the Zenith. As indicated already (page 8), *all Great Circles through the Zenith, also cross the Horizon at Right Angles*. Furthermore, we notice that the visible half of the Great Circle is always *halved* by the Zenith.

We can also place an indefinite number of Great Circles through the Pole Star, and these will cross the horizon at *varying* angles, but of course always at points diametrically opposite each other. The Pole Star halves only one of these Great Circles, namely that which crosses the horizon at the East and West (the centres of Figs. 7 and 9). By repeating these exercises with Great Circles, we recall to our memory the significance of a Great Circle. We become again fully aware of the fact that we, the observers, stand in the centre of each Great Circle.

North and South

The next step is that we place one Great Circle through both Zenith *and* the Pole Star. The best way of doing so is by facing East (e.g., the picture of Fig. 7). We stretch both our arms upwards towards the Zenith and then describe with our left arm that part of the Great Circle which passes down through the Pole Star, and with our right arm the other half. This Great Circle crosses the horizon vertically in the centre of Figs. 8 and 10, the left arm in Fig. 10, the right arm in Fig. 8. The centre of Fig. 10 we call North, the centre of Fig. 8 we call South.

We can also put it into these words. We face the Pole Star and place a Great Circle through the Zenith and Pole Star, by moving our vertically outstretched arm from the direction towards the Zenith, down through the Pole Star. This line crosses the Horizon in front of us in the North.

East and West

Now we face East, Fig. 7, so on our left is North, on our right is South. The centre of Fig. 7, in front of us, is called "East." The centre of Fig. 9, diametrically opposite, is called "West." This is not a very satisfactory definition of East and West, as it is arrived at so indirectly. We shall therefore attempt to find East and West in another way.

It may here be mentioned that the words North, South, East, West, are used to designate in each case a definite point on the Horizon and the direction in the sky marked by the Great Circle placed through this point and the Zenith. In other cases we shall refer to the Northern, Southern, Eastern and Western regions.

Let us once more recall how we established North and South. We placed a Great Circle through Zenith and Pole Star. This Great Circle stands vertically on the Horizon and divides the heavenly sphere into two equal parts. ***The finding of North and South is not based on the movement of stars, but on the linking together of two fixed points.*** One of them (the Zenith) does not depend on the movements of stars but on the stationary observer. The other one (the Pole Star) is a star, but owes its uniqueness to the fact that it does not participate in the movement of the stars but is practically immobile (see qualification made above).

This Great Circle linking up North (on the Horizon), Pole Star, Zenith, and South (on the Horizon) does not represent any movement of a star. It is not a line we can actually *see*, or which we can describe by following the movement of a star, but it *is* a line which we can describe if, as stationary observers, we link up the only two stationary elements in the sky (the Pole Star and our Zenith) with the other stationary element, the Horizon. The stationary character of all the components makes this Great Circle immensely suited for orientation at a given locality.

THE MERIDIAN CIRCLE

This becomes all the more evident if we now recall the star courses as described in Chapters V and VI. Small arcs in the South become greater and greater the more the points of rising and setting move towards the North, until even complete circles are reached. Finally we see in the "centre" of all these arcs and circles the Pole Star. From our discussion of the axis of the whole starry heavens, which is formed if we connect ourselves with the Pole Star, we can understand how the Pole Star, so to speak, "sums up" the whole starry heavens. By now adding our own Zenith to this fixed point, *we link together our position with the stars as a whole.* All star courses have one thing in common: they reach their highest point above the Horizon at this Great vertical Circle, extending from North to South via Pole Star and Zenith, and to the existence of which they actually have contributed so much. (The ***circumpolar stars*** reach both their highest and their lowest points here). All stars ***culminate*** in this Great Circle which is called the ***Meridian***. The reason for this name which is derived from the Latin word Medidies, (Midday) can only satisfactorily be explained in a later chapter on solar motions, when we will better appreciate what is meant by midday. The Meridian at any given locality is stationary and extends from South over Zenith and Pole Star to the North.

CELESTIAL EQUATOR CIRCLE

But let us once more return to the movements already discussed in Chapters V and VI. There we observed *movements* of stars, found more and more different movements as we proceeded, but were able, towards the end of Chapter VI, to harmonise all these movements again, to blend them into the majesty of one revolving starry heaven. We were able to single out of all star movements one star course (pages 13-14), which is unique, because it reveals itself as a Great Circle linking the centre of Fig. 7 with the centre of Fig. 9. On the Horizon these two points lie exactly halfway between North and South, diametrically opposite each other.

East and West are thus found to be the points on the Horizon at which those stars rise and set which describe an East-West Great Circle [***celestial equator***]. But this Great Circle does not pass through the Zenith, it does not stand vertically on the Horizon.

We face East (Fig. 7), and point with our right arm to the highest possible point of the East-West Great Circle (this point also lies on the Meridian). With our left arm we point towards the Pole Star, and it will be found that our two arms form a Right Angle.

In a later chapter we shall return to this aspect and shall be able to make it the starting point for very important further considerations. Here it may suffice to say this: the course of the stars rising in East and setting in West (centre of Figs. 7 and 9) is a Great Circle [***celestial equator***]. We have learned that the part of a Great Circle above the Horizon is a semi-circle, i.e. an equal part of such a star course lies above and another part below the Horizon. This is one reason why this star course is called ***the celestial Equator*** (from the Latin word for "equalise"). Further and still more elucidating reasons for this name can only be given in a later chapter when we have a clearer picture of the starry movements as a whole. It was however considered helpful to introduce the name Equator already, as the text will be more easily understandable.

While we established North and South by linking up stationary elements in the sky (Zenith and Pole Star) thru the Meridian, the East and West were established as the points on the Horizon which linked those stars which exactly follow the course of the Equator. **In other words, East and West are arrived at through movements of stars.**

Now, we have: Zenith and Horizon; Pole Star and celestial Equator; North and South; East and West. With these, we are able to orientate ourselves. Before we continue the discussion of what this new knowledge means for the study of the stars, we want however to make a beginning with the description of some stars and constellations, and shall do so in the next Chapter.

VIII. The Circumpolar Stars in the Northern Sky

Above the Northern horizon we find those stars which never rise nor set. They describe circular movements around the Pole Star, and are therefore called ***circumpolar stars***. We begin our study of individual stars and groups of stars in this region as it is an advantage to have the same stars before us whether we make our observation early or late, in Summer or in Winter.

The stars described in this chapter are always visible from our place of observation and further North. The further South we go the nearer stands the Pole Star to the Horizon and naturally the area of the circumpolar stars becomes smaller. One star of those mentioned below will soon become invisible in more Southerly regions on account of its sinking below the Northern horizon. It is VEGA in LYRA which on the Movable Planisphere actually disappears, the Planisphere having been designed for the latitude of Northern Switzerland.

Since very ancient times the observers of the heavens have linked certain stars into groups or pictures. Such groups are called ***Constellations***. The configurations of most of these constellations are about 5,000 years old, although different times and different countries have chosen different names. In E. Walter Maunder's "Astronomy without a Telescope" - (A guide to the Constellations, and Introduction to the Study of the Heavens with the Unassisted sight) - (London, W. Thacker & Co., 1904) - more details can be found about the origin of the constellations. This book can be a great help for the study of the stars, but it will be seen that for its full perusal a rather thorough study of astronomy is required. On the other hand the section "Constellation Studies" can be highly recommended in conjunction with Chapters VIII and X.

A hint as to the method of star observation may be useful. In a really clear night, we can be overwhelmed by the mass of stars we see in the sky and the feeling might arise in us that we shall never be able to find our way in the sky, that we shall never be able to tell the different constellations apart. We must therefore aim at the knowledge of certain "landmarks" first and commit to our memory their character, their "gesture". The sky is in perpetual movement around us and a constellation may look totally different in its position near the Zenith from when it stands near the Horizon. It is of great help if we learn to recognise a constellation by its "gesture" rather than by trying to pick out the individual stars of which it is composed. The number of stars visible at any one moment depends largely on the observer's eyesight and on the clearness of the sky. But if we know the gesture in addition to the position of a constellation relative to other constellations we shall find it easier to orientate ourselves. The "gesture" is represented by the line linking up the main stars of a constellation.

NORTHERN (CIRCUMPOLAR) STARS

Let us now look towards the Northern sky (especially that part which is above Fig. 10). As a time, we chose midwinter about midnight. [Note: for earlier observation times, 'back up' the rotation -- i.e., turn the diagram clockwise 90°. See above on using the planisphere to show varous dates and times.].

Most astronomical writers agree that there is one constellation which immediately catches the eye of the observer and which is known to almost everyone. At midnight about Christmas it is found halfway between Northeast on the Horizon and the Zenith. It is the ***GREAT BEAR***, (Latin: URSA MAJOR; German: GROSSER BAR).[1] The ***PLOUGH***, or ***BIG DIPPER*** is a well-known *portion* of this constellation, technically an 'asterism.'

Figure 12 - The Great Bear with the Big Dipper group

[Northern horizon, in late October @ Sunset (23h Sid. Time)]

---

1 The Latin and German names will be given in addition to the English to enable the readers to use also other star maps and particularly the Movable Planisphere.

The most familiar part of it is the group of seven stars connected by thick lines. It is pictured here in this upright position as it appears so at the time of our observation. In the course of the night it ascends higher and higher until it stands almost in the Zenith, to begin its descent in the morning hours.

The different successive positions can of course be watched on the Movable Planisphere by turning it to the right in the direction of the arrows.

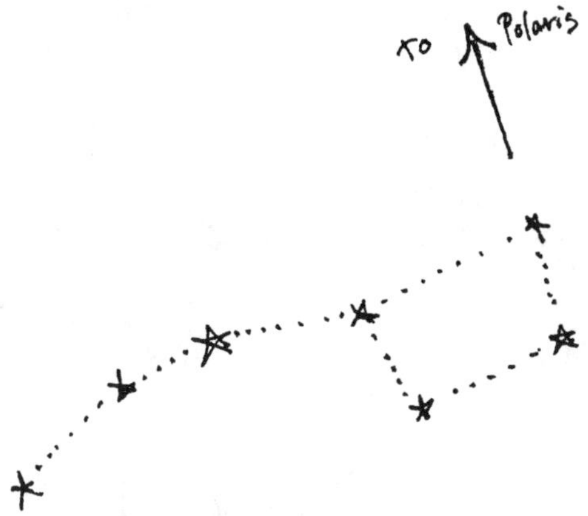

Figure 13: The "Gesture" of the Big Dipper

Fig. 13 indicates the "gesture" of the ***Big Dipper***. It is a valuable exercise to describe it in many different positions so that one really experiences the direction in which the "tail" points whatever the position of the constellation.

Let us now draw a line from ***the two "pointer" stars*** in the centre of the GREAT BEAR (in Fig. 12) in the direction of the dotted line, more or less parallel to the horizon. We come soon to a bright star which stands comparatively alone. It can be seen that this star is the one which we have recognized above the centre of Fig. 10 (page 11) as being practically stationary. We also see that it stands vertically above North on the Horizon, or in other words on the Meridian, the Great Circle extending from North to South via the Zenith. It is the ***POLE STAR*** (STELLA POLARIS, POLAR STERN), belonging to the constellation of the *Little* ***Bear*** which will be described later on.

We should try to grow accustomed to the angle our arm forms with the Horizon when pointing to the POLE STAR. For Chicago or Denver it is about 42°, while for Austin, Texas 32.5° [for Birmingham, England, the angle is 52.5°, while for Aberdeen, Scotland, it is little more than 57°.] It is a considerable help for our orientation in the sky if we know more or less accurately the angle at which to point to the POLE STAR, especially as this measure is important as well, for our orientation in the Southern sky as we saw at the end of the last chapter, when linking Pole and Equator.

We now continue the path from the GREAT BEAR to the POLE STAR on to the left (towards Northwest) and meet an easily recognisable constellation, ***CASSIOPEIA*** (or the LADY ON HER CHAIR or THRONE).

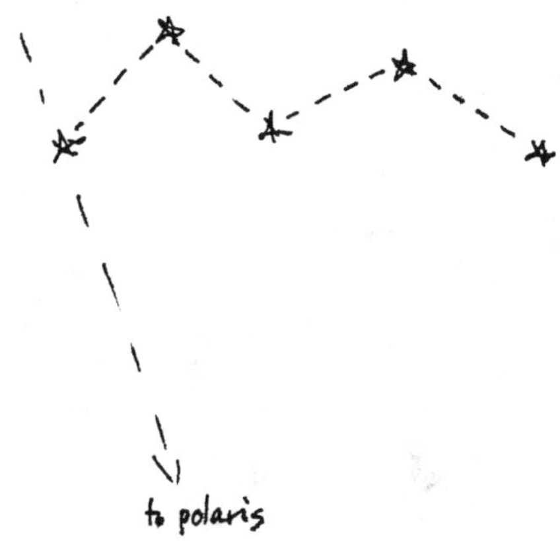

Figure 14: CASSIOPEIA

It actually stands a little higher in the Northwestern sky than the GREAT BEAR stands in the Northeastern, but the distance of both constellations is about equal. It will always be found opposite the GREAT BEAR if we take the POLE STAR as the centre. When following its course in the sky we do well to remember the "gesture" of CASSIOPEIA reminding us of a "W".

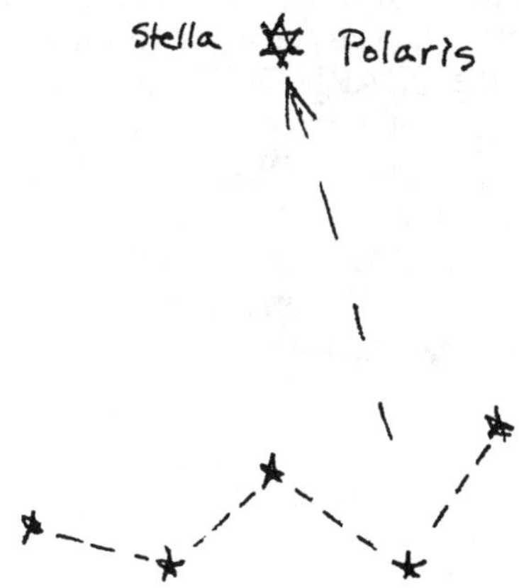

Figure 15: The "gesture" of CASSIOPEIA

Naturally this "W" will stand sometimes on its head, sometimes on its side.

Later, when studying the constellations of the Southern sky, we shall deal with the constellations further to the West (which we see on the Planisphere very close to the right margin); here, we want to confine ourselves to the circumpolar stars. Once more, we recall the almost horizontal path linking the

GREAT BEAR and CASSIOPEIA with the POLE STAR in the centre. If we now draw a line from the POLE STAR to the Zenith we meet a constellation which is easily recognisable because the region between it and the POLE STAR is fairly void of stars. A bright yellow star, as a matter of fact one of the very brightest in the Northern sky, is surrounded by four other stars.

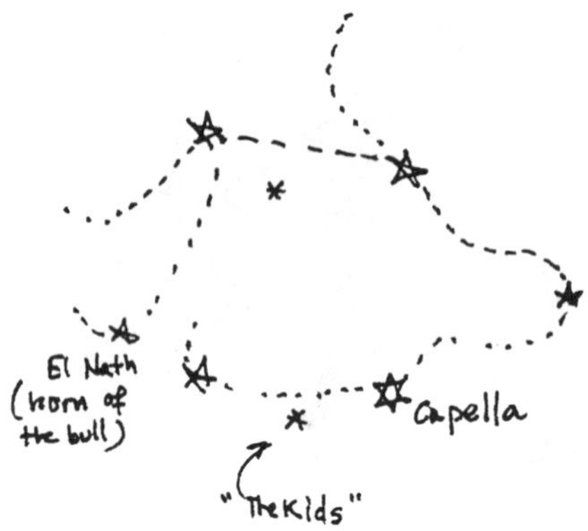

Figure 16: the Charioteer : AURIGA

The constellation is called the ***CHARIOTEER*** (AURIGA, FUHRMANN). The bright yellow star is ***CAPELLA***. The "gesture" of this constellation is a trapezium with CAPELLA standing on the side nearest to the POLE STAR. It will depend on the position of the observer whether he can see the constellation corresponding to the CHARIOTEER and CAPELLA. This stands at the end of a line drawn from the CHARIOTEER through the POLE STAR to a point slightly to the West of South on the Horizon. The further northwards the observer stands the greater are the chances of this constellation being visible even in its lowest position near the Horizon.

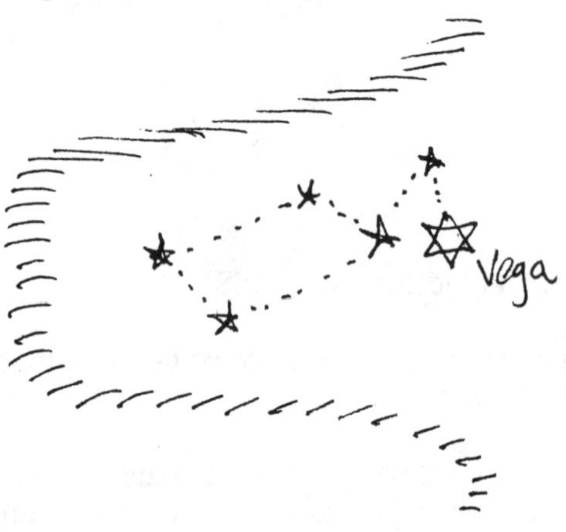

Figure 17: The Lyre (Lyra)

How this constellation received the name of ***LYRA*** (LEIER) can readily be understood. The bright steel-blue star in LYRA is ***VEGA***. It is of the same brightness as CAPELLA in the CHARIOTEER.

The Four Circumpolar Constellations

A mighty cross is formed by these four constellations from which we shall take our bearings to find five more.

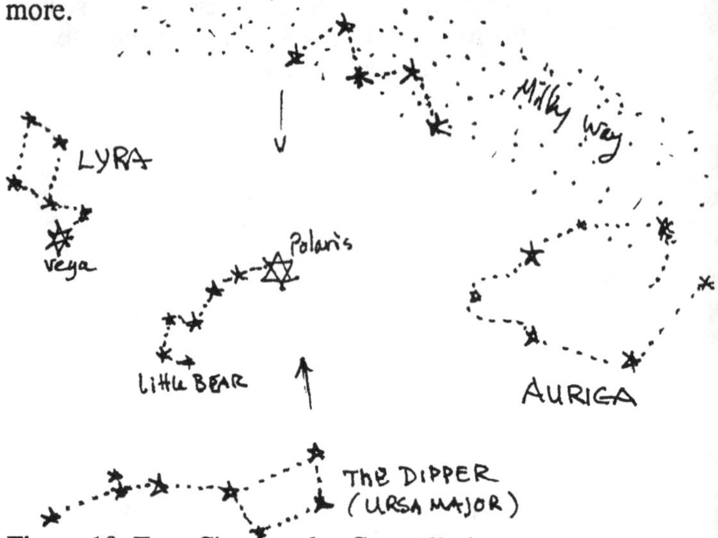

Figure 18: Four Circumpolar Constellations

CASSIOPEIA is easily recognized as situated in the Milky Way, which at Christmas time about midnight spans the heavens from Northwest to Southeast. Close to the Western horizon is a remarkable constellation, which is more easily recognized when it stands higher in the sky, e.g. in the Northwestern sky earlier in the evening.

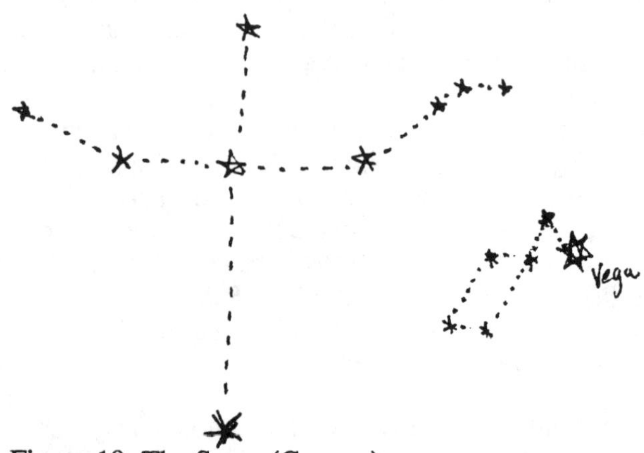

Figure 19: The Swan (Cygnus)
(setting on western horizon in November, sunset)

This constellation is called the ***SWAN*** (CYGNUS, SCHWAN). It stands close to LYRA and is the Milky Way. The bright star nearest to CASSIOPEIA is called ***DENEB***. In any position we shall recognise the SWAN by the *great cross* formed by its stars.

Halfway between the SWAN and CASSIOPEIA we find ***CEPHEUS***, a not very conspicuous constellation.

Figure 20: Cepheus

Another constellation with a characteristically bright star lies in the Milky Way between CASSIOPEIA and the CHARIOTEER: -- ***PERSEUS*** with the star ***ALGOL***.

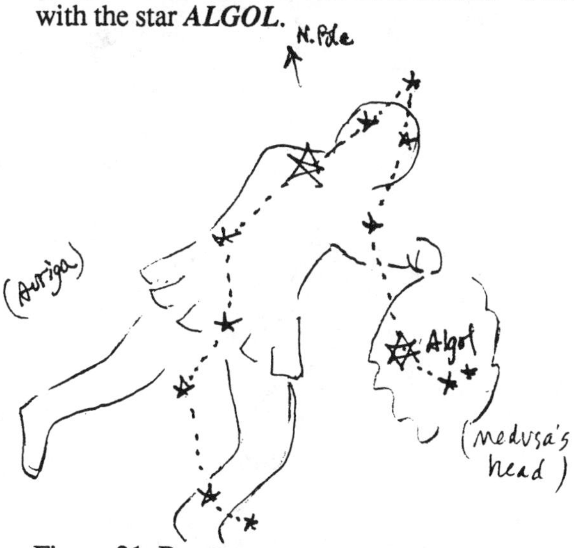

Figure 21: Perseus

This constellation is not always very easily identified as it looks so very different in its different positions. But from its location between CASSIOPEIA and the CHARIOTEER it can be found.

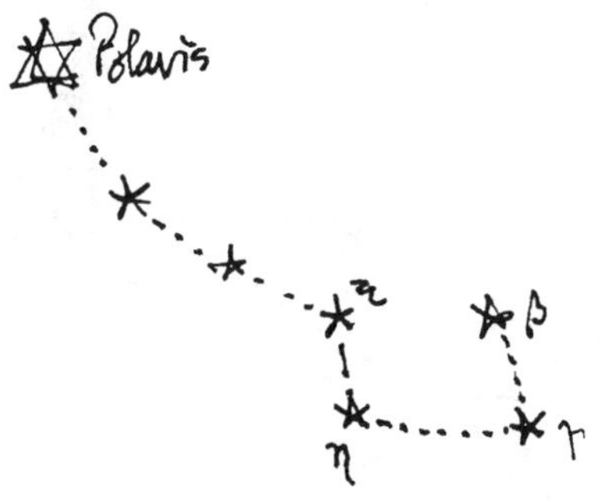

Figure 22: The Lesser Bear
[January, just over N. horizon]

Returning now to the POLE STAR we have seen that it belongs to the constellation of the LESSER BEAR (URSA MINOR; KLEINER BAR) the stars of which are not particularly bright.

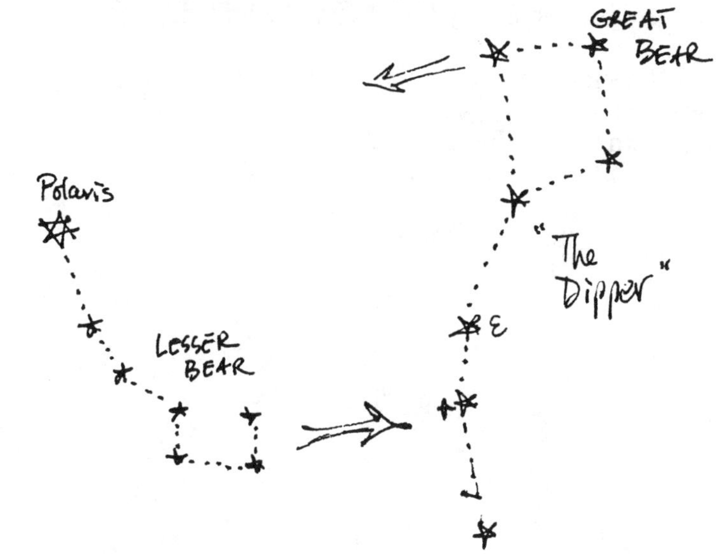

Figure 23: The relative positions of the Lesser Bear and the Greater Bear

Identification of the LESSER BEAR is much easier if one remembers that the "tail" is curved differently from that of the GREAT BEAR and that the whole constellation points towards the centre of the "tail" of the GREAT BEAR.

Figure 24: The Dragon (Draco)

In conclusion, we make acquaintance with a very long constellation. This is ***the DRAGON*** (***DRACO***, DRACHE), winding its way between the GREAT BEAR and LESSER BEAR and pointing with its head (at Christmas midnight) towards North on the Horizon.

This constellation is possibly the "Serpent" referred to in Job 26:13: "His hand hath formed the crooked serpent." In a Later chapter we shall see that in the folds of the DRAGON lies one of the most remarkable points of the whole sky (the Pole of the Ecliptic).

We sum up and unite the nine constellations into one picture (Fig. 25) which faces us as we look at the Northern sky at midnight at Christmas.

The Southern sky will offer much greater difficulties and some introductory remarks must be made before we can describe what we see there.

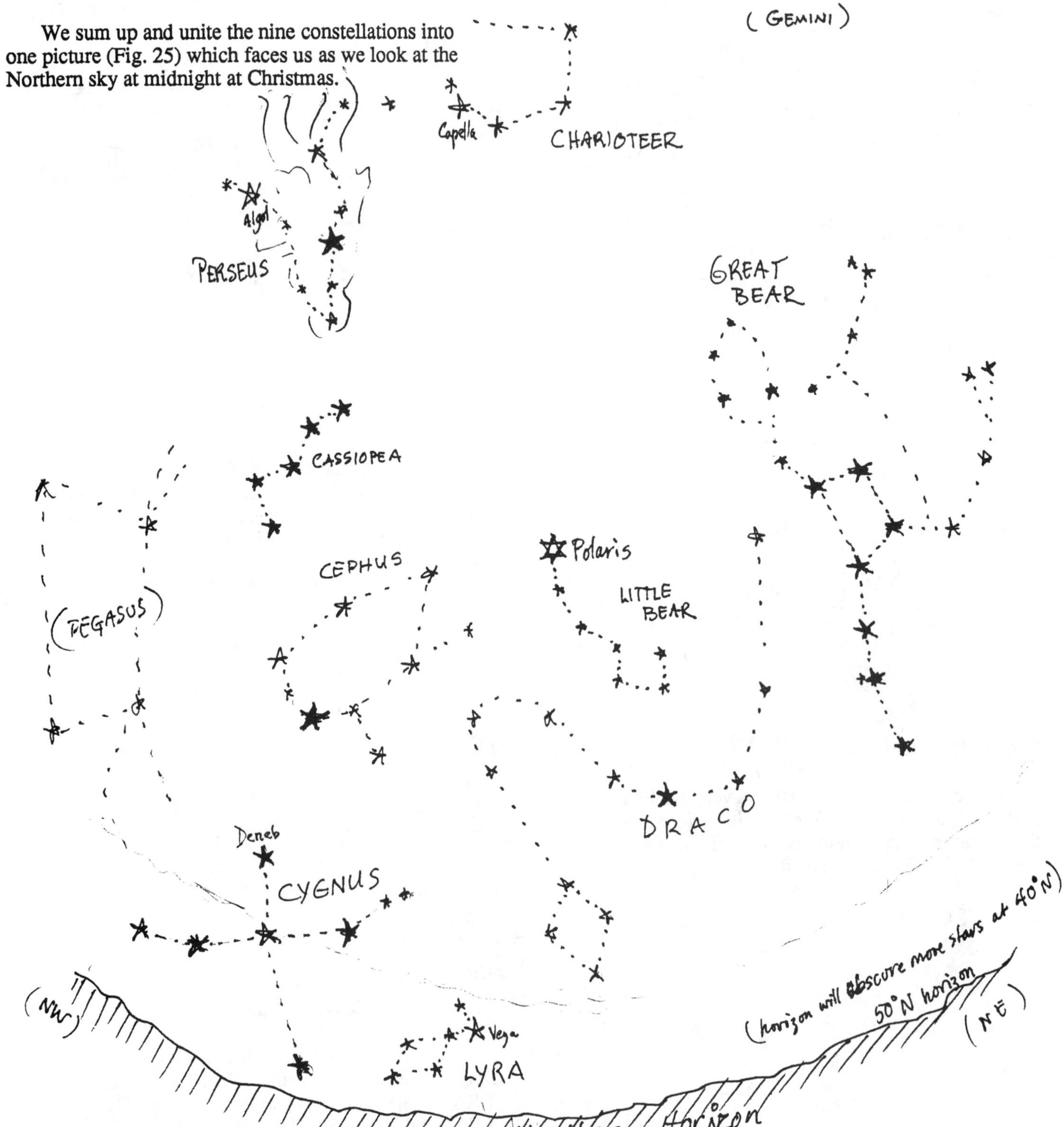

Figure 25: The Nine Major Circumpolar Constellations (midnight at Christmas) (~ 10PM Feb21 6h Sid.T.)

IX. The Study of the Southern Sky and the Further Use of the Movable Planisphere.

Introductory Remarks

In Chapter VIII we studied the main star constellations of the Northern sky. The "circumpolar" nature of all these stars (with the possible exception of one or two) enabled us to view the stars of the Northern sky in their totality, and the only real difficulty a beginner may experience is that a constellation at one time will stand "upright", while at another time it may stand "upside down". We remember e.g. CASSIOPEIA which, when it appears near the horizon, looks like a "W", but when standing overhead, near our Zenith, is not so easily recognisable. To overcome this difficulty, we suggested committing the "gestures" to memory. The fact remains that in a star-lit night we can generally see all these nine constellations above the Horizon. The only issue is their orientation, right or left, above or below the Pole Star. [Note: this 'cap' of the starry heavens has been compared to a watch face, where the four major circumpolar constellations move about the Pole star like the hands of a clock. This is the basis for a "Nocturnal," a device which tells time by the position of the 'pointer' stars in the Big Dipper.]

The Southern Sky

When we now turn towards the Southern sky we meet a totally different situation, which we ought to experience to the fullest extent. We must only be aware of the elementary fact that the Sun never stands in the North but always in the Southern sky with which we are going to deal now. Already at this stage, we can understand therefore that a firm knowledge of the star movements in the Southern sky will be a good foundation for the study of the Sun's movement later on.

The most remarkable difference between the Northern and Southern sky is that in the Southern sky we find no stationary star like the POLE STAR. Instead of such a point, however, we have *a line*. This is the Equator, the course of all those stars which rise due East and set due West. As the points of rising and setting are diametrically opposite each other and we therefore stand in the centre, this arc must be a Great Circle - hence we see it as a straight line. If we face East, and point with our left arm to the POLE STAR, and meanwhile keep our right arm at a right angle to our left arm, the right arm points towards the culminating point of the Equator which naturally lies on the Meridian (see Chapter VII) [and exactly over the South on the Horizon]. This is of great practical value for finding the position of the Equator. (In the picture of the Southern sky at midnight at Christmas (Fig. 27), the right-most star in the belt of ORION stands at the point towards which our right arm points.

The courses of all stars are parallel to each other. That means that a star course never crosses the Equator; it remains always either above or below. A star remaining below the Equator rises somewhere South of East, and sets somewhere between South and West. The points of rising and setting must, naturally, be the same distance from South. The arc described by such a star is shorter than that of the Equator and it will culminate lower. In other words, the nearer to South a star rises and sets, the less high it rises for its culmination (in the Meridian).

On the other hand, there are stars which rise between East and North and set between West and North. Such stars remain longer above the Horizon, and rise to greater heights.

All stars rising in or near East, or near South, differ in an important further point from the stars of the Northern sky inasmuch as they never stand "upside down". Near the Horizon, they will look as though "lying down", but we never see them reversed. This is a more qualitative difference, which should not, however, be ignored.

Most of what has been said so far in this chapter does not represent anything new. But we are now at the important point where we have to apply practically what so far has been achieved as a more theoretical knowledge of star movements. The readers are well advised once more before entering into the next chapter, to practise these exercises so that they become quite familiar with, and aware of, the parallel nature of all these movements. We shall seldom have the opportunity to watch the stars for hours at a time. We must therefore become able to describe with our arms the movement of any star. If we are clear as to the position of the Equator, we can visualise the past or future course of the star, without difficulties.

Further on the Planisphere

To those readers who have a movable Planisphere at their disposal, a few words of advice:

First of all Chapter IV should be read once more carefully so that it is clear that in the two dimensions of a map we see things differently from how they appear in the three dimensions of space around us. As soon as we use now the Planisphere as an aid for practical observation, we shall find out where on the Southern and Northern side of the map our Zenith is to be found. We shall also quickly become used to the fact that in reality East and West lie left and right of us when we face South, and not in front of us. - Our place of observation corresponds to the little brass rivet around which the Planisphere revolves.

If we want to "set" the Planisphere for a definite time, we must clearly realism that *the black scale* inside represents the hours of the day. 13^h is 1 pm, 14^h is 2 pm, etc., until we reach 24^h at midnight.

On *the white margin* we find the 365 days of the year. - To "set" the Planisphere we have to bring together the hour of the day with the day of the year.

For example: We want to "set" the Planisphere for the first picture of the next chapter: 24th December, 6 p.m.: 6 p.m. is 18h, which is on the right hand side of the black frame (Southern side). Next we turn the movable part of the Planisphere until 24th. December stands in the middle of the little slot, just at 18h. It is a useful habit to turn the map always in the direction of the arrows, as this movement represents that of the stars.

Armed with a renewed knowledge and experience of star movements described in previous chapters, and at least some of us with a Planisphere and a knowledge how to use it, we begin now with the study of the stars in the Southern sky.

X. The Constellations over the Southern Horizon.

We begin our observation of the stars in the Southern sky in the *early* evening hours of a midwinter night, let us say at 6 p.m. on Christmas Eve. This exact time is mentioned here so that readers can "set" their Planisphere accordingly. The picture described in Fig. 26 reveals itself before our eyes.

First we ascertain the position of the Equator by describing a Great Circle from East to West. In the Southwest it goes through MIRA in the constellation of the WHALE (CETUS: WALFISCH). MIRA (the Wonderful) is one of the many "variable" stars which change considerably in brightness.

In the corresponding position in the Southwest we find on the Equator the constellation of the EAGLE (AQUILA: ADLER) with the bright ALTAIR (or ATAIR) just above the Equator, the whole constellation standing in the Milky Way.

Thus having found the Equator we see from the map in Fig. 26 that in the Southwest the constellations of WATERMAN (AQUARIUS: WASSERMANN) and GOAT (CAPRICORN: STEINBOCK) stand below the Equator, but these constellations are not very distinguishable.

We now turn our attention to the constellations above the Equator and begin in the Northeast with the TWINS (GEMINI: ZWILLINGE) which have just risen. This constellation will be described when dealing with Fig. 27, when it stands in a more prominent position.

In the East the BULL (TAURUS: STIER) can be clearly seen. It is a great constellation full of interest. The two most easterly stars form the points of the horns, the orange-coloured star more or less in the centre is ALDEBARAN, the Follower, as it follows the PLEIADES, a star cluster slightly higher.

In the Southwest stand the RAM (ARIES: WIDDER). It is not a very bright constellation and only three stars are comparatively easily recognized. Above it stands the TRIANGLE adjoining the constellations described as belonging to the Northern sky.

Next come the FISHES (PISCES: FISCHE), the westerly one, which stands on the Equator, just culminating at this hour. This constellation is not very conspicuous either.

Above it extends over a considerable area in the Zenith-region PEGASUS, which passes over (towards East) into ANDROMEDA.

Two constellations stand high above the Equator in Southwest, but so high that they have been dealt with already in the Northern sky, namely SWAN and LYRA.

Close to the Horizon, about to set, we find HERCULES, touching (at the end furthest away from the Horizon) the DRAGON in the Northern sky.

o O o O o

We pass on to the picture which reveals itself six hours later and see in Fig. 27 the Southern sky at midnight during midwinter and ascertain at first the position of those stars which help us to locate the

Equator. MIRA in the WHALE has moved from Southeast to Southwest and where MIRA stood at 6 p.m. we see now a bright star PROCYON in the LESSER DOG (CANIS MINOR: KLEINER HUND).

The most brilliant and dominating constellation in the southern sky is now ORION. There is hardly a constellation to which more allusions have been made in literature from the oldest times right down to our present age. The "belt" of ORION stands on the Equator and at this time of the night we should use the opportunity to describe the Equator again and again with our arms so that we get really and truly familiar with its position. ORION is easily picked out in the sky, as some of its stars are very bright. The one on the Northeast corner, orange in colour, is called BETELGEUSE; the one diametrically opposite on the Southwestern corner is RIGEL, the brightest of them all.

By means of ORION we have no difficulty in locating a number of other constellations. BETELGEUSE and PROCYON form together with SIRIUS in the GREAT DOG (CANIS MAJOR: GROSSER HUND) and equilateral triangle. - The bluish-white SIRIUS stands below the Equator and is therefore visible for a considerably shorter time only. SIRIUS has the distinction of being by far the brightest star in the whole sky.

The constellation of the HARE (LEPUS: HASE) at the foot of ORION is not very easily distinguished.

The beauty of the midwinter sky at Christmas by no means ends with these constellations. Above the Equator we find in the Western sky the constellations of FISHES, RAM and BULL, (enumerating them in their sequence from West towards South). They have been described in connection with Fig. 26.

Above ORION towards the East, just above PROCYON in the LESSER DOG, stands the conspicuous constellation of the TWINS

(GEMINI: ZWILLINGE) with CASTOR (higher) and POLLUX (lower) at the Eastern end of this distinctive oblong.

From the TWINS we move towards the Eastern horizon and meet the CRAB (CANCER: KREBS) with PRAESEPE in its centre. This constellation can only be found easily on a clear night. Much brighter is the next constellation which has just risen: LION, (LEO: LOWE). It is easily recognized by the "sickle" like Western part, two stars of which deserve a special name. The one is whitish in colour and stands nearer the Equator, REGULUS, the other - shining in yellowish light - slightly above (i.e. in the present position to the left), ALGIEBA.

In a later Chapter we shall realism more fully the great significance of this Christmas-midnight picture. At hardly any other time do we see such a magnificent display of stars in the Southern sky. Many stars have been named and the reader might at first be bewildered; but if he will take his time and first carefully place the Equator by describing the Great Circle from East through PROCYON, the belt of ORION, MIRA to West, he separates the "above" from the "below". Above ORION, in perfect balance, shine the TWINS and the BULL. Then further apart CRAB and RAM, the nearest to the horizon, LION and the FISHES.

o O o O o

Again six hours later, we observe the Southern sky at 6 a.m. at Christmas (Fig. 28). To find the Equator we turn our attention to Southwest and see PROCYON in the LESSER DOG where at midnight MIRA stood. The Southeasterly part of the Equator passes through the large constellation of the VIRGIN, (VIRGO: JUNGFRAU). The brightest star of this picture, SPICA, stands slightly below the Equator.

Below the Equator, slightly to the East of the Meridian, stands the small but fairly conspicuous constellation of the CROW (CORVUS: RABE). Right across the Southern sky, close to the Horizon, extends HYDRA, in the WATERSNAKE, the longest constellation in the sky, the "head" of which reaches almost to the CRAB.

If we now turn our attention to the constellations above the Equator we see in the Northwestern Horizon the BULL just disappearing, followed by the TWINS, CRAB and LION, which latter has just recently passed the Meridian. In the Southeast, fairly high up, we see BOOTES with the brightly shining ARCTURUS. In the East stands HERCULES which we met in the evening picture of Fig. 26 in the West. Between BOOTES and HERCULES we see the NORTHERN CROWN (CORONA BOREALIS: KRONE).

o O o O o

In order to complete the series of pictures we can see in the Southern sky we show in Fig. 29 the Southern sky at midnight during midsummer, although during winter never, neither early nor late, do we see the constellations in these positions. The Equator goes through the EAGLE in the Southeast and OPHIUCHUS, the SERPENTHOLDER, in the South and crosses the SERPENT in Southwest. Near the Western Horizon the Equator passes through VIRGIN.

Below the Equator stand the rather inconspicuous constellations of GOAT (CAPRICORN: STEINBOCK) in Southeast, ARCHER (SAGITTARIUS: SCHUETZE) in South, SCORPION (with the bright ANTARES) and the SCALES (LIBRA, WAGE) in Southwest. The close proximity of all these to the Horizon makes them all the more difficult to identify.

In the East spreads PEGASUS. In the Southeast above the EAGLE stands the SWAN, high above OPHIUCHUS HERCULES, followed towards West by the CROWN and BOOTES with ARCTURUS.

All these constellations are already known to us and we have only to note here their position differing so widely from those at other times.

o O o O o

We could not possibly attempt to describe in detail every constellation and its position, but we have chosen the more important ones. The knowledge of their positions and movements will have made alive the more abstract considerations of earlier chapters. Those readers not satisfied with the number and choice of constellations and stars will have no difficulty in obtaining further information in Star-Atlases or astronomical textbooks. For us going through this Course a number of questions begins to formulate itself as to the "why" of all these movements. How can we reconcile the regularity of the starry heavens with the appearance of one and the same constellation at so many different places at different times?

To answer such questions will be the task of the next chapters.

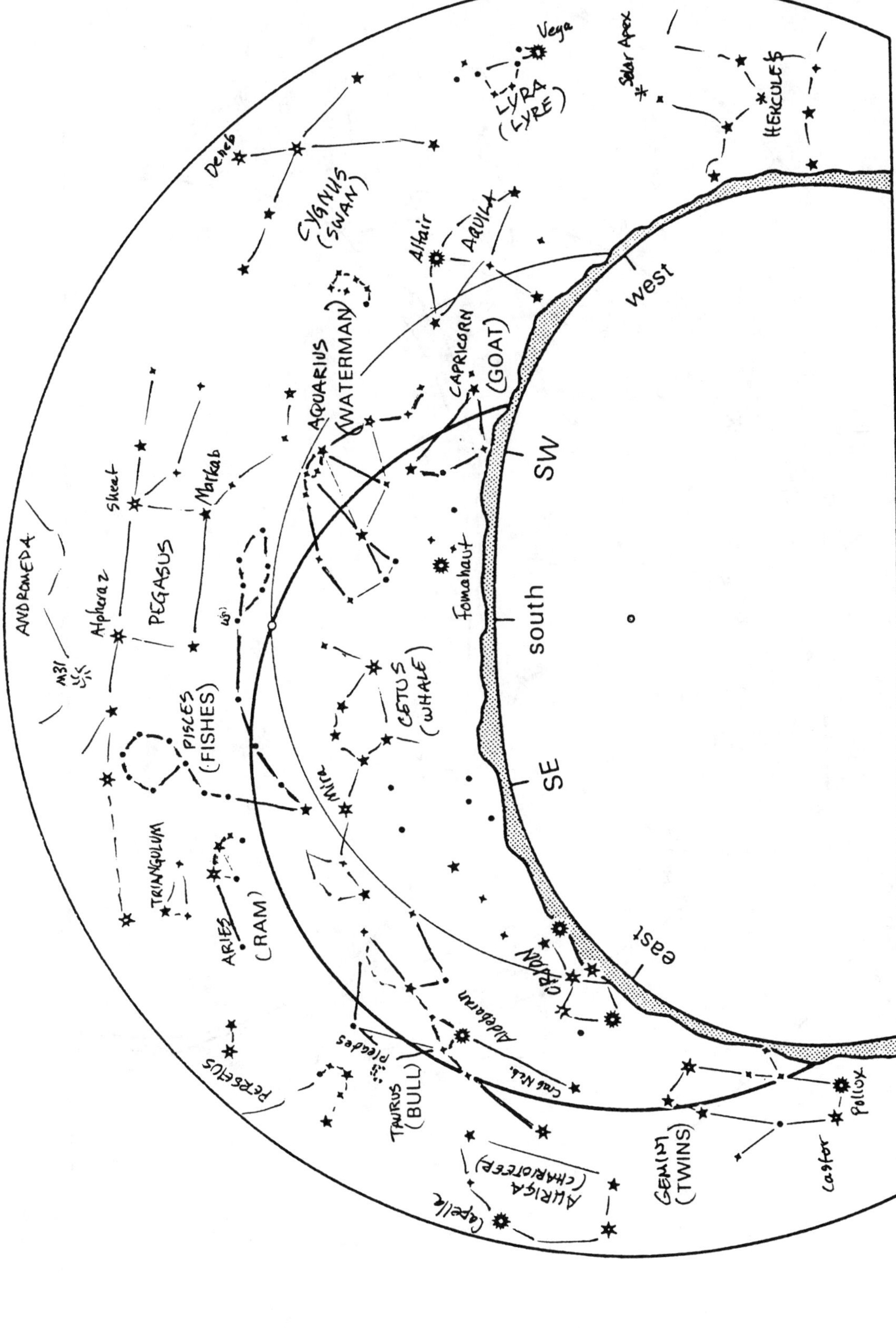

Figure 26: The mid-winter evening heavens with the Zodiac as it appears March 21st @ 12:00, June 21st @ 6:00 am, Sept. 23 @ 24:00, **Dec. 22nd @ 18:00** (0h sidereal time; zodiac in intermediate position).

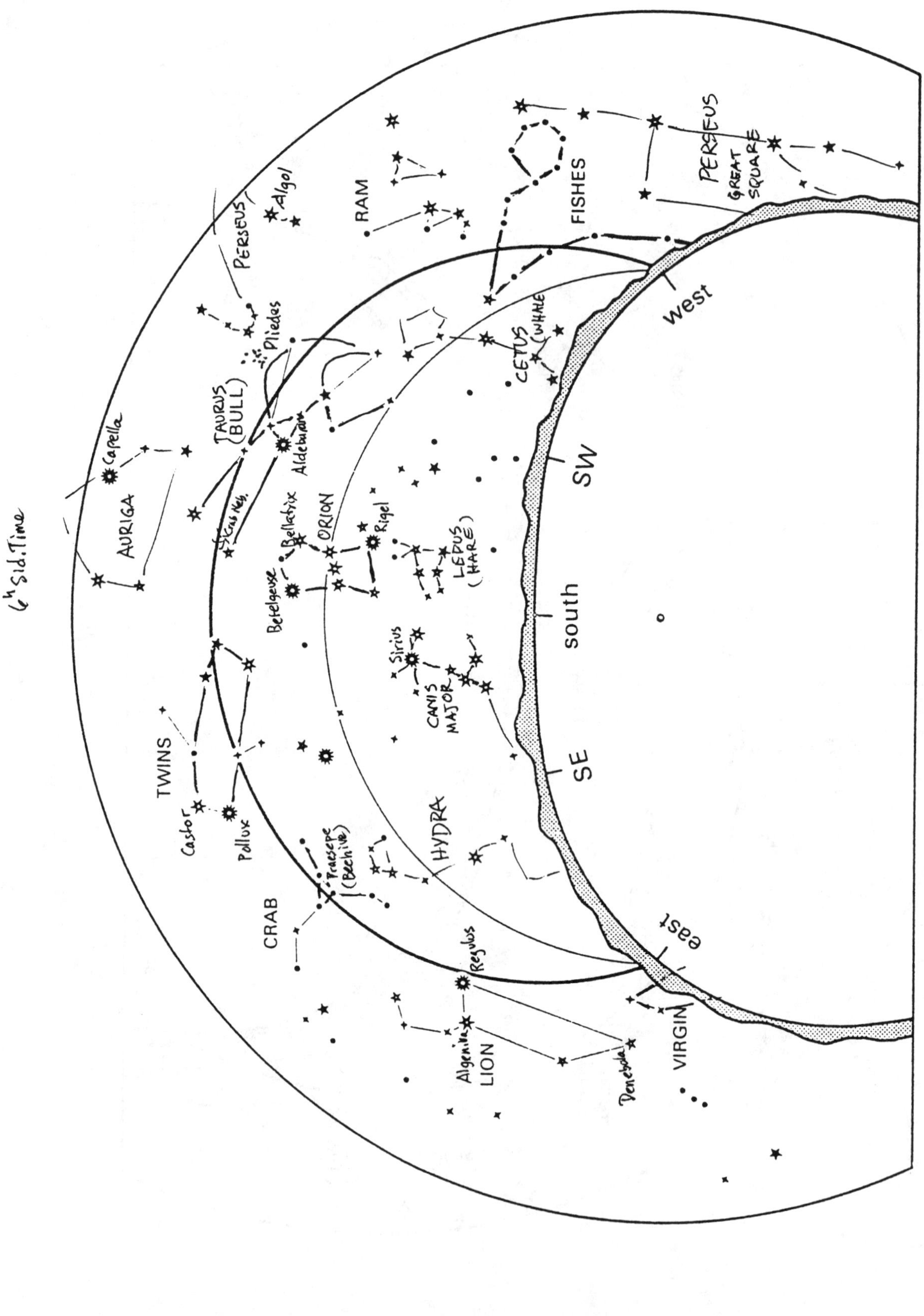

Figure 27: The Spring (evening) Zodiac constellations as they appear: **March 21st @ 18:00**, June 21st @ 12:00 am, Sept. 23 @ 6:00, Dec. 22nd @ 24:00 (6^h sidereal time; zodiac in intermediate position).

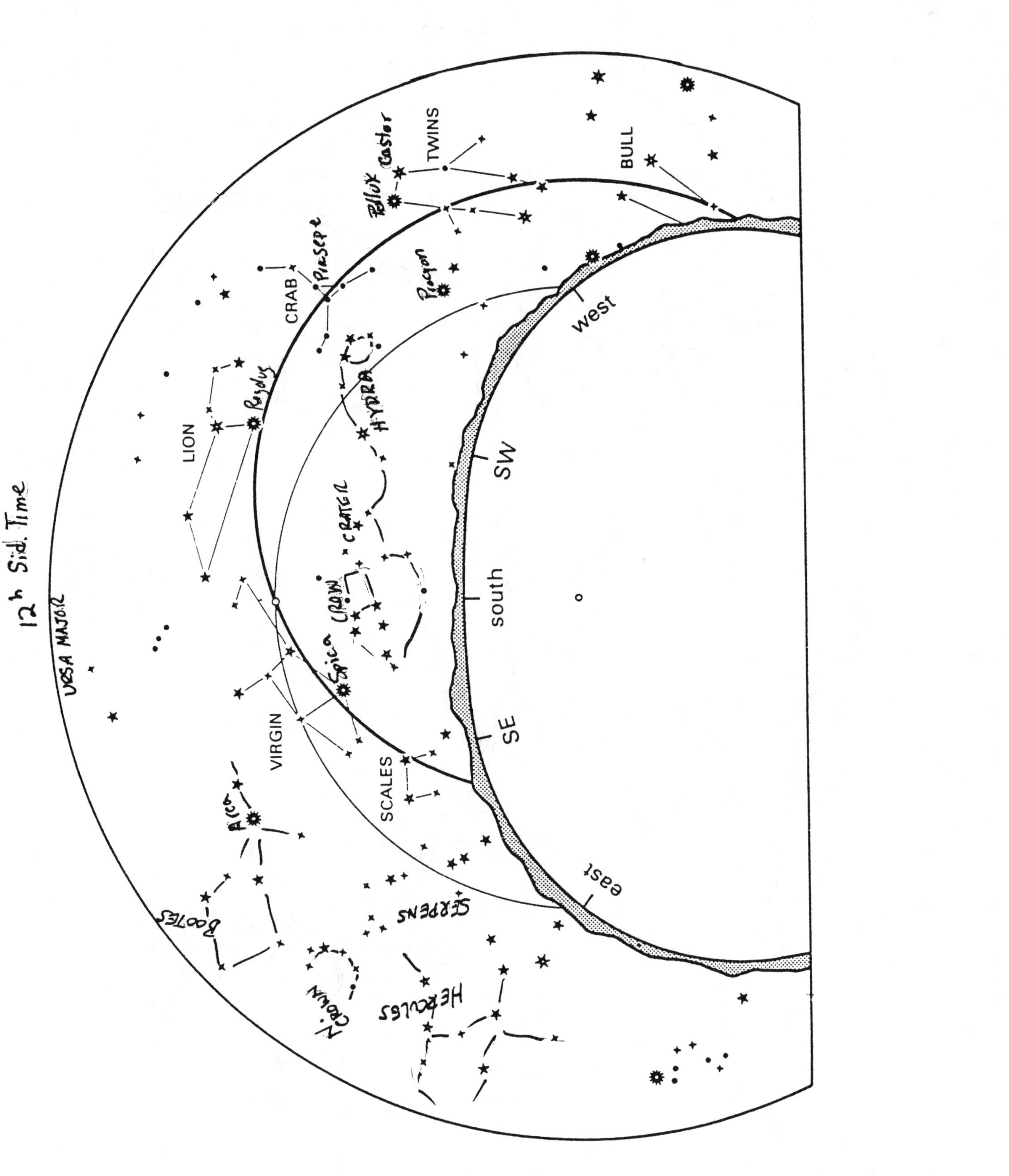

Figure 28: The Summer (evening) constellations as they appear: March 21st @ 24:00, **June 21st @ 18:00 am,** Sept. 23 @ 12:00, Dec. 22nd @ 6:00 (12^h sidereal time; zodiac in intermediate position).

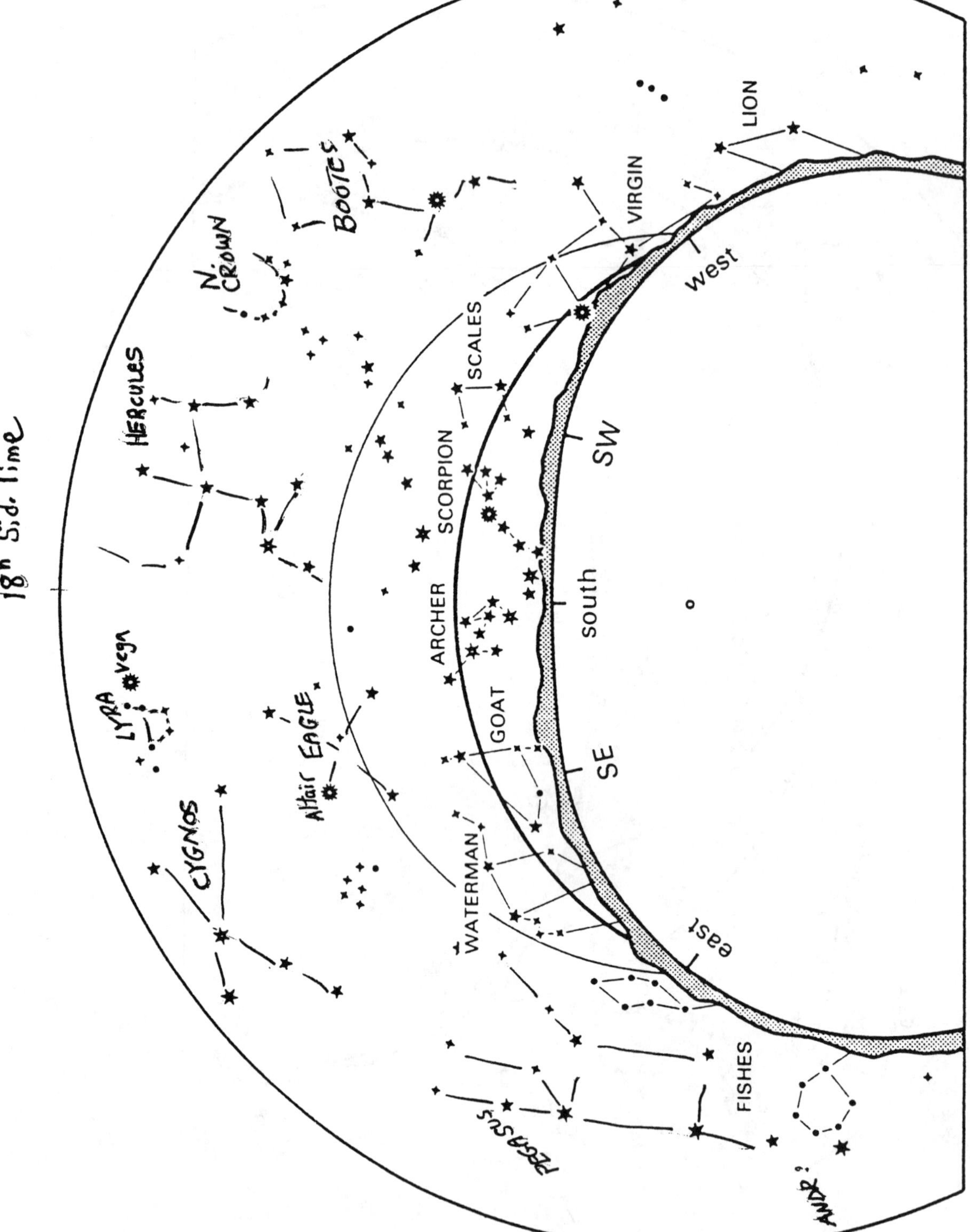

Figure 29: The Autumn (evening) constellations as they appear: March 21st @ 6:00, June 21st @ 24:00 am, **Sept. 23 @ 18:00,** Dec. 22nd @ 12:00 (18^h sidereal time; zodiac in intermediate position).

Directions

1. Cut out the star circle on page 33 and paste it on a similar sized piece of cardboard.
2. Cut out the semi-circle below and also paste it on a piece of cardboard or stiff paper.
3. With a pin, pierce a hole in the center of the star circle and the point indicated on the semi-circle. Place the semi-circle over the star circle and attach with a clip or paper rivet.
4. You should now be able to rotate the semi-circle on the star circle. This is your completed Planisphere.

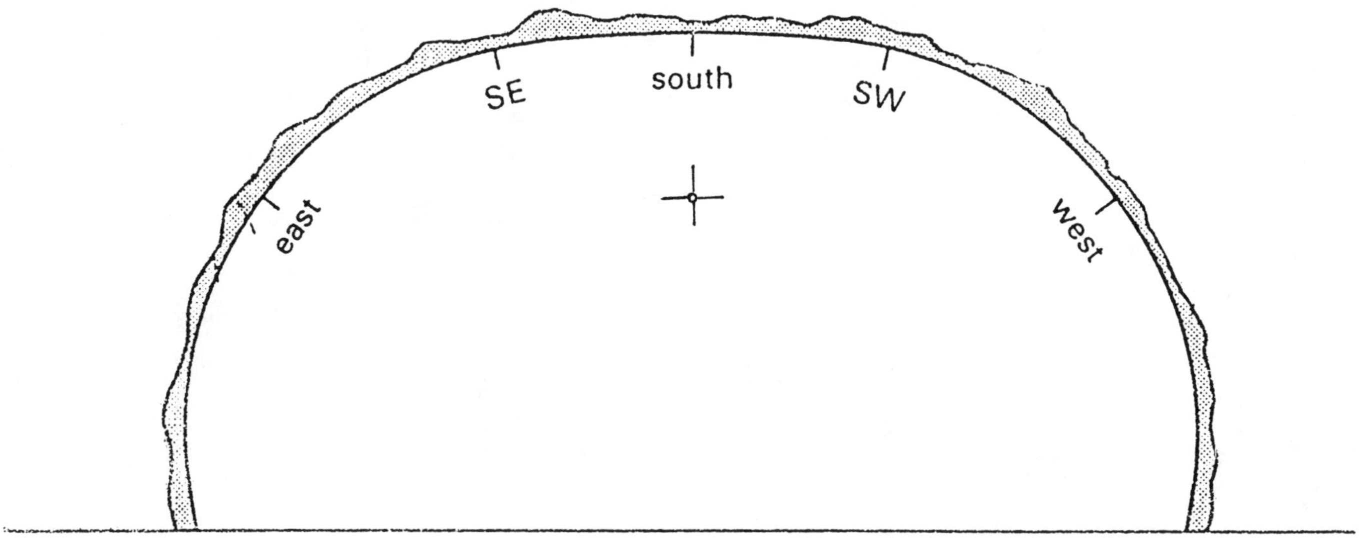

12h

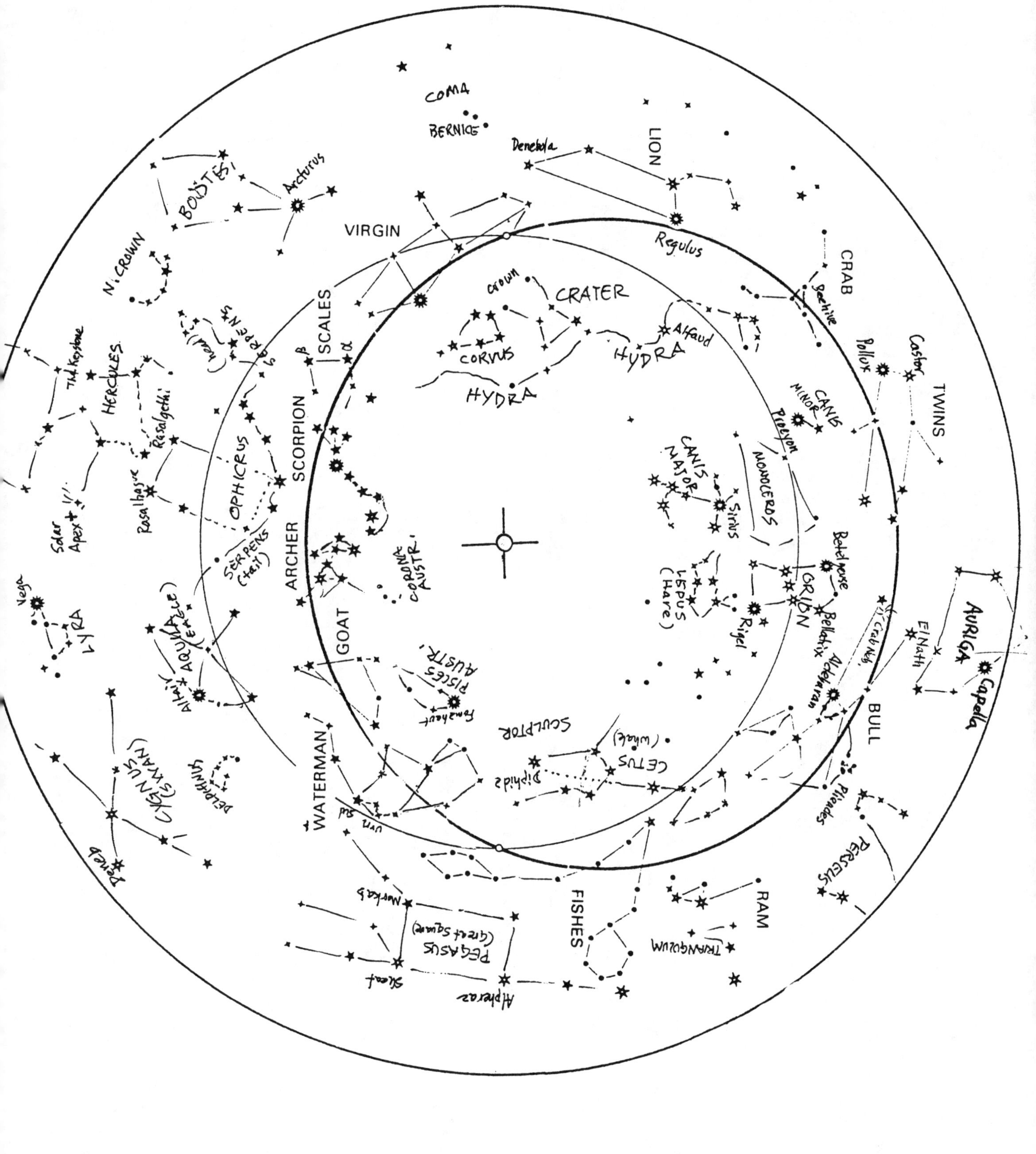

COMA
BERNICE
Denebola
LION
Regulus
BOOTES
Arcturus
VIRGIN
CRAB
Beehive
N. CROWN
SERPENS
SCALES
Crown
CRATER
HYDRA
Alfard
CORVUS
HERCULES
The Keystone
Rasalgethi
Pollux
Castor
TWINS
CANIS MINOR
Procyon
SCORPION
OPHICRUS
CANIS MAJOR
MONOCEROS
Sirius
Rasalhague
Solar Apex
ARCHER
SERPENS (tail)
CORONA AUSTR.
Betelgeuse
ORION
LEPUS (Hare)
Rigel
Bellatrix
Vega
LYRA
AQUILA (EAGLE)
Altair
GOAT
PISCES AUSTR.
Fomalhaut
AURIGA
El Nath
Capella
Crab Neb.
Aldebaran
BULL
SCULPTOR
CETUS (whale)
Diphda
CYGNUS (SWAN)
DELPHINUS
WATERMAN
Pleiades
PERSEUS
Deneb
FISHES
RAM
Markab
PEGASUS (Great Square)
TRIANGULUM
Scheat
Alpheraz